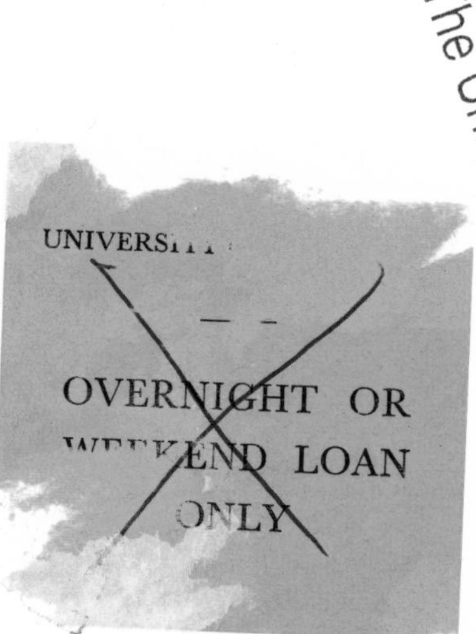
UNIVERSITY
OVERNIGHT OR
WEEKEND LOAN
ONLY

DEVELOPMENTAL AND CELL BIOLOGY SERIES

EDITORS

M. ABERCROMBIE   D. R. NEWTH

J. G. TORREY

CYTODIFFERENTIATION IN PLANTS

CYTODIFFERENTIATION IN PLANTS

Xylogenesis as a Model System

LORIN W. ROBERTS

Professor of Botany, University of Idaho

FOREWORD BY

JOHN G. TORREY

Professor of Botany and Director of the Cabot Foundation, Harvard University

CAMBRIDGE UNIVERSITY PRESS

CAMBRIDGE

LONDON · NEW YORK · MELBOURNE

Published by the Syndics of the Cambridge University Press
The Pitt Building, Trumpington Street, Cambridge CB2 1RP
Bentley House, 200 Euston Road, London NW1 2DB
32 East 57th Street, New York, NY 10022, USA
296 Beaconsfield Parade, Middle Park, Melbourne 3206, Australia

Library of Congress catalogue card number: 75-10041

ISBN: 0 521 20804 1

First published 1976

Printed in the United States of America
by Vail–Ballou Press, Inc.,
Binghamton, New York

Library of Congress Cataloging in Publication Data

Roberts, Lorin Watson.
Cytodifferentiation in plants.
(Developmental and cell biology series)
Bibliography: p.
Includes indexes.
1. Plant cells and tissues. 2. Cell differentiation. 3. Xylem. I. Title.
QK725.R72 581.8'761 75-10041
ISBN 0-521-20804-1

Contents

Foreword

The modern problems in understanding cytodifferentiation lie in that middle ground between the refined and elaborately detailed classical studies of plant anatomy performed at the histological and light microscope level and the ever-increasingly complex and integrated molecular mechanisms and circuitries worked out by the biochemist. Cytodifferentiation is a cellular problem best seen expressed in a multicellular tissue system but most easily studied in a not-yet-discovered 'ideal' cell suspension system where fully synchronized units pass from A to Z in concert and *in toto* upon a given signal.

It is timely for the field of study to attempt to examine from as many different viewpoints as possible the differentiation of a single cell type to see how far we have come in developing a comprehensive understanding of how a cell differentiates and to put up a model for studying other systems.

The selection of cell differentiation in xylem is the obvious and perhaps the only choice. Analysis of cytodifferentiation in tracheary elements in the vascular tissues of higher plants goes back at least to the work of Crüger in 1855 and in the past century-and-a-quarter more attention has been paid to this distinctive cellular type than to any other single differentiated plant element. As a tissue component, the tracheid or vessel element is one of the most diagnostic cell types, characterizing as it does the 'Tracheophytes' or vascular plants, including the ancient fossil psilophytes, the lycopods, or club mosses, the sphenopsids or horsetails, the living ferns, the gymnosperms and the angiosperms.

As the subject for analysis the tracheary element is the cell of choice. One of the earliest visible (under the light microscope) events in the differentiation course is the deposition inside the primary cell wall of a thick secondary cell wall usually formed in precise if bizarre patterns of annular, reticulate or pitted design. This secondary wall of cellulose is quickly encrusted with lignin which conveys special staining properties to an already highly specialized birefringent (under polarizing lenses) cellular feature. When the secondary wall is complete and the end walls dissolved away in the areas of the perforation plate, if present, the whole cell contents autolyze, leaving an empty hollow element which in an integrated tissue system forms a conducting pipe for the transport of water and solutes.

The differentiation of the conducting elements of the xylem, a feature in the morphogenesis of every organ of the vascular plant, occurs in a highly organized way, assuring an integrated pathway for water flow from cells of the ultimate root tips throughout all of the plant body to the ultimate endings in leaves whence the water is lost from the plant. It is a process making possible the tremendous heights of our tallest living trees and is associated with the successful conquest of the land by the present vegetation. Clearly, the whole plant exercises rigid control on the prepatterning for vascular differentiation as well as the cytodifferentiation of all of the connected elements which comprise the xylem.

We are a long way from understanding the ultimate detailed processes whereby a cell produced by cell division in the apical meristem of a shoot or root undergoes directional enlargement, shifts its nuclear and cytoplasmic activities toward the formation of new wall materials and finally passes into programmed cell death, assuring a free path for specific and highly efficient transport within the multicellular organism. It is the nature of the control processes and the events of the differentiation itself which form the content of this book.

Professor Roberts has brought together in a comprehensive and thoughtful monograph the existing literature which has developed, particularly over the last two or three decades, in the middle ground of the cell biology of xylem cytodifferentiation. To be best appreciated his treatment must be thought of in the context of the extensive and elegant work on xylem in the existing anatomical literature on the one hand and the increasingly articulate and intricate information being unravelled at the molecular level by the developmental biologists, both plant and animal, on the other. In reviewing carefully and critically the experimental work on xylogenesis, the preceding cell cycle events and the associated biochemical processes especially of macromolecular changes, Professor Roberts has placed cell botanists interested in cytodifferentiation in his debt.

Petersham, Massachusetts John G. Torrey
September 1974

Preface

Much of the research in the area of plant cytodifferentiation has centered on the formation of tracheary elements. Because of their unique appearance, these cells are readily detected in plant tissues; moreover, under experimentally controlled in-vitro conditions, the induction of cytodifferentiation of these cells has been achieved in several systems. Certain basic questions occurred during the gestation period of this book. How can we devise future experiments for a more precise control and regulation of the initiation of xylem differentiation? Can xylogenesis be employed by developmental botanists as a model system for the study of the cytodifferentiation of other types of plant cells? Can we relate the molecular aspects of the cell cycle to the onset of xylogenesis? The answers to these questions would be of paramount importance for the cytodifferentiation of other plant cell types.

I share with my fellow developmental botanists a defensive posture. There remains, for example, a wide discrepancy between our knowledge about the biochemistry of growth regulators and their physiological and developmental significance. Developmental studies have invariably lacked the precision characteristic of the physical sciences. Our field, however, is likely to remain largely a descriptive science until we have discovered factual explanations concerning the modes of action of the basic plant hormones. We need not apologize for our hypotheses and speculations; the foundations of scientific advancement have been the dreams and imagination of scientists.

L. W. R.

Acknowledgments

Most of this book was written in the spring of 1974 during my appointment as a Maria Moors Cabot Research Fellow at Harvard University. I am most grateful to John G. Torrey for his generous assistance. Without his warm encouragement and constructive criticism this book would not have been written. I deeply appreciate and fondly remember the many courtesies extended me by the faculty and staff of the Harvard Forest and the Maria Moors Cabot Foundation for Botanical Research. I am particularly grateful to Randy Landgren for many of the illustrations, and to Peter Barlow for his valuable comments on the cell cycle. Thanks are also due to Richard Zobel, Graeme Berlyn, David Dobbins, Tsvi Sachs and Tore Timell for suggestions concerning cytodifferentiation. The generosity of friends who provided me with original plates and negatives for many of the illustrations is acknowledged with the figures. Finally, I am grateful to my wife Florence for her editorial suggestions.

Abbreviations

ADP, ATP	adenosine diphosphate, adenosine triphosphate
AMO-1618	2′-isopropyl-4′-(trimethylammonium chloride)-5′-methylphenyl piperidine carboxylate
A-state	cell cycle period similar to G_0; metabolism not directed toward DNA replication
B-phase	cell actively engaged in cell cycling
cAMP	adenosine 3′,5′-cyclic monophosphate
8-Br-cAMP	8-bromoadenosine 3′,5′-cyclic monophosphate
CCC	β-chloroethyltrimethylammonium chloride (chlorocholine chloride; Cycocel)
CDS	cytodifferentiation sequence
CFl	2-chloro-9-hydroxyfluorene-9-carboxylic acid (chlorfluorenol; IT-3299; IT-3456)
C value	relative amount of DNA per nucleus; normal diploid cell after DNA synthesis and before mitosis has 4C amount of DNA
2,4-D	2,4-dichlorophenoxyacetic acid
DAB	3,3′-diaminobenzidine
dgt	single gene mutant (*diageotropica*) of tomato requiring exogenous ethylene for normal development
DNA	deoxyribonucleic acid
DPX-1840	3,3a-dihydro-2-(*p*-methoxyphenyl)-8*H*-pyrazolo-(5,1-a)-isoindol-8-one
E.C.	enzyme classification number
ER	endoplasmic reticulum
FUdR	5-fluoro-deoxyuridine
G_0	postulated period of cell in a nonproliferating condition; similar to A-state
G_1	period of cell cycle following mitosis and prior to DNA synthesis
G_2	period of cell cycle following DNA synthesis and prior to mitosis
GA	gibberellic acid; gibberellin A_3; GA_3
IAA	indole-3-acetic acid

IAASA	sucrose–agar medium containing IAA
M	period of cell cycle during mitosis
mRNA	messenger ribonucleic acid
NAA	α-naphthaleneacetic acid
NAD	nicotinamide adenine dinucleotide
NDP	nucleoside diphosphate; N = uridine, adenosine, thymidine, cytidine or guanosine
P_{730}	far-red absorbing form of phytochrome pigment
PAL	L-phenylalanine ammonia-lyase
PESIGS	mnemonic device referring to the criteria for the determination of chemical critical variables in a CDS; Parallel variation, Excision, Substitution, Isolation, Generality and Specificity
rDNA	ribosomal deoxyribonucleic acid; genes coding for ribosomal ribonucleic acid
R_f	ratio of the distance the solute moves divided by the distance the solvent moves from the point of origin on the chromatogram; e.g., a value of 0.5 would indicate that a given solute had moved one-half of the distance from the loading site to the solvent front.
RNA	ribonucleic acid
rRNA	ribosomal ribonucleic acid
S	period of the cell cycle during DNA synthesis
S2, S4	strains of *Acer pseudoplatanus* callus
S_1, S_2, S_3	secondary wall layers deposited successively on the inner surface of the primary wall during secondary xylem element differentiation
25 S-[^{3}H]rRNA	25 Svedberg tritiated ribosomal ribonucleic acid fraction employed in hybridization studies with DNA
SA	agar medium containing sucrose
TE	tracheary element
TH-6241	5-methyl-7-chloro-4-ethoxycarbonylmethoxy-1,2,3-benzothiadiazole
TIBA	2,3,5-triiodobenzoic acid
tRNA	transfer ribonucleic acid
UDP	uridine diphosphate
VFN8	isogenic normal tomato; parent of *dgt* mutant

1

Cytodifferentiation in perspective

The term 'cytodifferentiation' is confusing because it implies both the state of a given cell as well as a developmental process. Cells are differentiated with respect to other cells when they differ; cells undergo differentiation when they pass from one state to another (Heslop-Harrison, 1967). There are, however, no 'undifferentiated' cells without genetic programs, and terminally differentiated cells are the products of precursor cells with unique genetic programs (Holtzer & Abbott, 1968). How dividing cells are genetically reprogrammed to produce predictable changes in the metabolism between parent and daughter cells (Holtzer & Bischoff, 1970) is the basic problem. Gross (1968) has defined differentiation as the sum of the processes by which the acquisition of specific metabolic competences (or the loss thereof) distinguishes daughter cells from each other or from the parental cell. At the molecular level we may consider two cells differentiated with respect to each other, if, while they contain the same genome, the pattern of protein synthesized by the two cells is different (Jacob & Monod, 1963). Cytodifferentiation starts with the transformation of the metabolism of a cell into a unique pathway that results in the appearance of new and different biochemical properties. These biochemical properties, in turn, lead to certain structural specializations. Determination is the process by which a cell becomes restricted to a new pathway of specialization, and in this discussion the term 'point of determination' will refer to that transient period during which cytodifferentiation is initiated. Cytodifferentiation may be reversible, and there are numerous examples of the dedifferentiation of specialized cells (Gautheret, 1966). The relative ease of callus formation from the culture of a variety of specialized plant tissues indicates the genetic totipotency of mature plant cells (Heslop-Harrison, 1967).

In order to avoid ambiguity, the term 'cytodifferentiation sequence' (CDS) will refer to the entire developmental process from the point of determination to the mature, fully differentiated xylem element. Two basic types of tracheary elements occur in the xylem of the Tracheophyta; tracheids and vessel members (Esau, 1965*a*). The CDS of these water-conducting elements includes the following sequential stages of development: cell origination, cell enlargement, secondary wall deposition and lignification, and wall lysis and cell autolysis (Torrey, Fosket & Hepler, 1971). The period of time devoted to each of these stages in a given cell may vary, and thus the CDS is not of a uniform time period for all differentiating tracheary elements. The extent and direction of cell elongation may vary considerably from one tracheary element to another. In cultured explants of lettuce pith the minimum for the CDS is approximately four days (Dalessandro, 1973*a*), whereas a minimum of seven days is required for the same process in cortical explants of pea root (Phillips & Torrey, 1973). A similar sequence in the secondary xylem may require several weeks for completion (Skene, 1969). Experiments by Siebers (1971*a*) have suggested that interfascicular cells may pass the point of determination and be arrested from further development until they receive the proper hormonal cue to a resumption of the CDS.

The CDS involves time in two senses (Wright, 1973). First, it involves a series of sequential processes. Second, at any given point during the CDS several metabolic systems are functioning simultaneously, and each of these systems requires time for its respective development, i.e., synthesis and activation of enzymes and formation of enzyme substrates. For example, a conspicuous feature of the cytodifferentiation of tracheary elements is the deposition of the distinctive secondary wall. Let us assume that following the excision and culture of a particular explant on a xylogenic medium, the point of determination for xylem differentiation was at sixty hours and secondary wall deposition commenced at eighty hours. Does the cell prepare for the biosynthesis of the necessary enzymes and substrates for secondary wall formation at sixty hours or perhaps at seventy-five hours? Is gene programming for the entire process completed at the time of determination, or does the process go a step at a time with the metabolic consequences of one pathway initiating reactions that lead to the following stages? This is one of the basic questions concerning the cytodifferentiation of higher plant cells.

The terminology of Wright (1973) is useful in examining the parameters of cytodifferentiation. At any given point during the CDS certain critical variables or 'limiting events' are specifically necessary in order for the CDS to proceed. There are also 'controlling events' that are nonspecialized functions required for normal cell metabolism, but which are only indirectly related to the CDS. Probably our most pressing problem concerns the means of discovering the critical variables at the point of determination, and at subsequent stages, for a given CDS. A similar dichotomy of general and specific functions has been given by other workers (Holtzer & Abbott, 1968; Ephrussi, 1972). This review examines tracheary element formation as a model system for the study of the parameters involved in the unfolding of the CDS, and it is hoped that some of the concepts presented may prove useful in framing studies concerning the CDS for diverse types of plant cells. The major emphasis is placed on internal control mechanisms, since environmental factors appear to modify xylem differentiation rather than act as critical variables in the CDS. This is not meant to be an exhaustive review of the current literature on xylem differentiation. The observations of several workers on the appearance of tracheary elements under a wide variety of different experimental conditions were purposely omitted since their findings were not pertinent to the problems under discussion. Following a brief historical introduction, the roles of various hormones in primary xylem differentiation are reviewed. In a more exploratory vein, an examination is made of the possible relationships between the cell cycle and cytodifferentiation. The hormonal requirements for the cytodifferentiation of cambial derivatives is examined in view of the unique problems associated with the developmental physiology of the secondary xylem. Ultrastructural evidence for organelle functions in cytodifferentiation is discussed, followed by chapters on nutritional and environmental factors as controlling variables. A final chapter is devoted to the applications of certain chemical inhibitors to cytodifferentiation studies.

For additional information on various aspects of xylem differentiation the reader is directed to the publications of Esau (1965*b*), Halperin (1969), Lipetz (1970), Torrey *et al.* (1971), Zimmermann & Brown (1971), Kozlowski (1971) and Gamaley (1972). Most of the early literature on the physiology of xylem differentiation was reviewed by the author in 1969.

2

Historical survey of xylem differentiation studies

The earliest observations on the experimental induction of tracheary element differentiation came primarily from three different types of investigations: the regeneration of vascular tissue in wounded stems and leaves, the induction of cytodifferentiation in plant tissue cultures and the formation of secondary xylem elements following cambial reactivation. Relatively little work has been done on the hormonal regulation of primary vascular tissue differentiation either in the root or shoot (Torrey, 1966).

The cytodifferentiation of vascular tissues induced by wounding was first described by Vöchting (1892), and subsequently studied by Simon (1908*a*), Freundlich (1908), Kaan Albest (1934), Jost (1940, 1942) and Sinnott & Bloch (1944, 1945). Jost (1942) found that the application of indole-3-acetic acid (IAA) to wounded leaves resulted in the differentiation of xylem strands within the leaf mesophyll. The pattern of secondary wall thickenings that extended from cell to cell indicated that newly formed tracheary elements were in 'continuity' with previously differentiated xylem cells (Fig. 1), and the suggestion was made that the wall configurations were prepatterned by band-like thickenings of the cytoplasm (Sinnott & Bloch, 1944, 1945). Previously Crüger (1855) and Dippel (1867) had made similar suggestions. Although several ultrastructural studies have failed to confirm these early observations (see Chapter 6), the problem is still under investigation by Goosen-De Roo (1973*a, b, c*). According to early investigators tracheary elements around a stem wound were formed in a strictly basipetal manner (Fig. 2). The observation that removing the leaves and buds distal to a wound site greatly reduced the extent of cytodifferentiation (Kaan Albest, 1934) suggested to Jacobs (1952) that diffusible auxin from these distal sources was the limiting factor for the initiation of cytodifferentiation. Subsequent reports from Jacobs' laboratory (1954, 1956; Jacobs & Morrow, 1957) on the quantitation of the auxin-induced xylogenic response in stem wounds of *Coleus* provided the main impetus for our current research efforts on auxin as one of the key factors in the cytodifferentiation of tracheary elements. As a result of these studies, Jacobs (1959) formulated a set of criteria for the determination of chemical critical variables in a cytodifferentiation sequence. These criteria are designated by the mnemonic 'PESIGS', which represents Parallel variation, Excision, Substitution, Isolation, Generality and Specificity. If a given chemical limits a particular cytodifferentiation process, then the following criteria should be met:

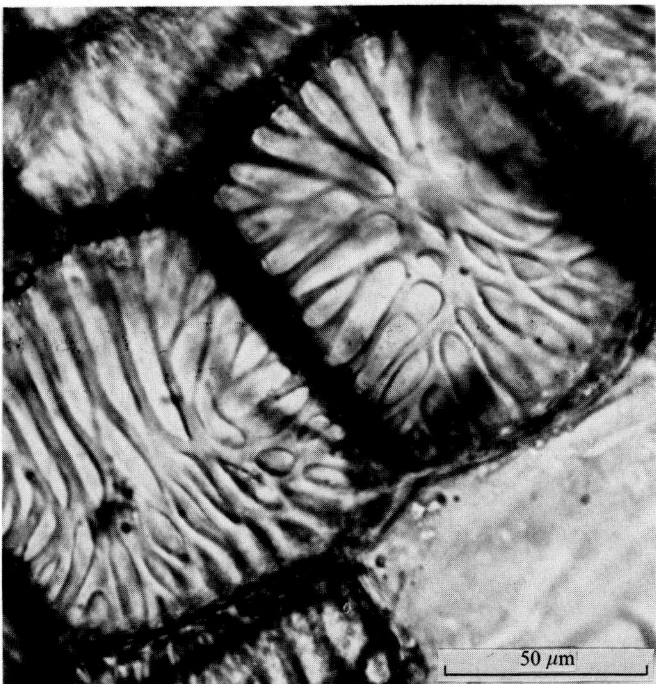

Fig. 1. Contiguous tracheary elements usually display a remarkable similarity in the patterns of secondary wall thickenings, which suggests that pattern determination during cytodifferentiation can be passed from cell to cell. The two tracheary elements in the photomicrograph differentiated in an explant of lettuce pith parenchyma cultured on a xylogenic medium.

1. The extent of cytodifferentiation *in vivo* should vary directly with the amount of the chemical.

2. The removal of the chemical, e.g., by excision of the source, selection of genetic mutants, or by the use of specific inhibitors, should demonstrate a subsequent absence of the cytodifferentiation.

3. Substitution of the chemical for the normal in-vivo source of the chemical in the organisms should restore the cytodifferentiation process.

4. Attempts should be made to isolate and, as near as possible, define, the parameters of the system in terms of biochemistry and cytology.

5. Generality of the results may be demonstrated by showing that these criteria are valid in different families and in various plant organs.

6. The specificity of the chemical for the process should demonstrate that other naturally occurring chemicals have no such effect on the cytodifferentiation process (Jacobs, 1959).

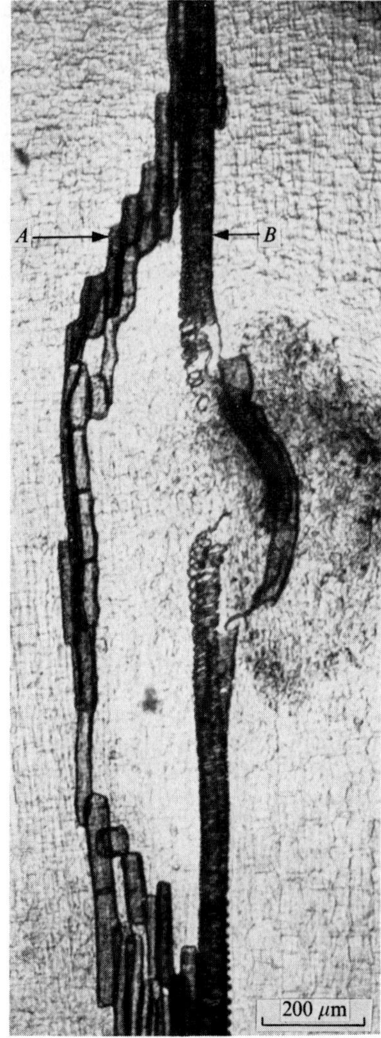

Fig. 2. Regeneration of a strand of tracheary elements (*A*) around a ruptured vascular bundle (*B*) in a *Coleus* stem. Vascular bundle was interrupted by a puncture wound produced with a glass microneedle. (From Roberts & Fosket, *Bot. Gaz.* **123,** pp. 247–54, 1962; courtesy of University of Chicago Press.)

It is ironic that many developmental physiologists have employed the rationale of these criteria in the design of their experiments without being aware of Jacobs' pioneer efforts in this area.

Numerous attempts have been made to control, by chemical manipulation of various media, the initiation of tracheary element differentiation in cultured tissues. The production of xylem elements in wound-induced callus, at the base of stem cuttings, was described by Simon (1908*b*). Interest in the possibility of hormonal regulation started with the discovery that auxin applied to decapitated plants resulted in the formation of masses of 'wound xylem' (Blum, 1941; Whiting & Murray, 1946; Beal, 1951). Bud initiation in callus (Gautheret, 1942) and bud grafting into callus (Camus, 1949) were both found to induce xylogenesis in the adjacent tissues, and presumably the effect was due to the release of auxin (Camus, 1949). Camus (1949) observed that chicory (*Cichorium intybus*) buds grafted to explants of endive (*Cichorium endivia*) root tissue grown in culture induced cytodifferentiation. Buds grafted into the phloem induced vascular strand formation basipetally toward the cambium of the stock, whereas buds grafted into xylem parenchyma produced isolated zones of cytodifferentiation. Wetmore & Sorokin (1955) found that *Syringa* buds grafted to stem callus of *Syringa* induced xylogenesis, and a similar xylogenic effect was achieved by replacing the bud with a mixture of IAA and sucrose in agar.

Later Wetmore & Rier (1963), using the same technique, demonstrated that the position and type of vascular tissue induced to form in callus was dependent on the relative concentrations of the exogenous auxin and sucrose applied to the apical wedge of the callus. Relatively low sucrose concentrations favored xylem formation, high sucrose levels favored phloem, and intermediate levels produced both xylem and phloem in vascularized nodules of meristematic cells. Clutter (1960) reported that the application of IAA to explants of tobacco pith by means of a micropipette inserted into the tissue induced cytodifferentiation. Several workers have demonstrated that exogenous sugar is a necessity for xylogenesis (Wetmore & Sorokin, 1955; Fosket & Roberts, 1964), and various sugars may modify the response (Ball, 1955; Jeffs & Northcote, 1967). Jeffs & Northcote (1967) reported that organized xylem in vascular nodules was induced in explants of bean callus only in the presence of a disaccharide containing an α-glycosyl radical at the nonreducing end of the molecule; all other sugars tested induced the formation of scattered tracheary elements.

Other workers also have attempted to relate a specific carbohydrate requirement to the induction of xylem differentiation (Karstens & de Meester-Manger Cats, 1960; van Lith-Vroom, Gottenbos & Karstens, 1960; Rier & Beslow, 1967; Beslow & Rier, 1969). However, some of the carbohydrate effects on xylogenesis may be due to variations in water potential (Doley & Leyton, 1970).

The interaction of several groups of plant hormones in cytodifferentiation has been demonstrated by the use of in-vitro cultures. Xylem differentiation was initiated only in the presence of both exogenous auxin and cytokinin in a large number of agar and suspension cultures (Bergmann, 1964; Torrey, 1968; Fosket & Torrey, 1969; Torrey & Fosket, 1970; Dalessandro & Roberts, 1971; Phillips & Torrey, 1973). Gibberellic acid demonstrated synergism in the cytodifferentiation response produced by combinations of auxin and cytokinin (Gautheret, 1961*a*; Roberts & Fosket, 1966; Dalessandro, 1973*a*). Siebers (1971*b*) indicated that a combination of gibberellic acid and kinetin was more effective than auxin in stimulating the interfascicular cambium to produce xylem elements. There is some evidence that adenosine 3′, 5′-cyclic monophosphate (cAMP) may play a role in cytodifferentiation (Basile, Wood & Braun, 1973), although this may be an adenine effect (Ferré, 1971). Recent evidence suggests that ethylene, at extremely low concentrations, plays an important role in xylem differentiation (Zobel, 1973). A possibility exists that abscisic acid is a factor in the cessation of cambial activity and differentiation (Hess & Sachs, 1972). The literature on the experimental induction of tracheary element differentiation in cultured plant tissues has been reviewed by Gautheret (1966), Halperin (1969), Roberts (1969), Torrey (1966, 1971) and Torrey *et al.* (1971).

Hartig (1853) observed that the seasonal reactivation of the cambium was associated with expanding buds. During the early research on IAA as a natural plant auxin, the concept was developed that IAA produced by actively growing buds was transported basipetally, and was the hormonal stimulus for initiating cell division in the dormant cambium (Zimmerman, 1936; Avery, Burkholder & Creighton, 1937; Söding, 1937). Although IAA appears to be an important factor, other equally important factors are involved in cambial reactivation. The cambium exhibited a seasonal responsiveness to hormonal applications (Reinders-Gouwentak, 1941; Zajaczkowski, 1973), which may reflect the presence of inhibitors such as abscisic acid. Auxin-induced cambial activity was greatly stimulated by exogenous gibberellic acid (Wareing, 1958; Wareing, Haney & Digby, 1964; Digby & Wareing, 1966), and the production of gibberellins by the root system in the spring may be a limiting factor in cambial reactivation (Lavender *et al.*, 1973). Bradley & Crane (1957) found that spur shoots of apricot sprayed with gibberellic acid showed a stimulation of cambial activity and enhanced production of secondary xylem. Sorokin, Mathur & Thimann (1962) examined the effects of kinetin, in the presence or absence of IAA or 2,4-dichlorophenoxyacetic acid (2,4-D), on segments of etiolated pea epicotyl. The auxins alone stimulated cambial activity with the production of slightly abnormal secondary xylem. Kinetin treatment, however, either in the presence or absence of added auxin, produced considerably more cambial proliferation with the formation of apparently normal secondary xylem.

The role of pressure in secondary xylem differentiation (Brown & Sax, 1962; Brown, 1964) suggests that stress-induced ethylene production is involved in cytodifferentiation. Experiments by Evert and his co-workers (1967, 1972) involved the surgical isolation of circular patches of bark that remained *in situ* on the sides of the trees. Secondary xylem differentiation beneath the bark patches isolated during dormancy was nearly completely inhibited, and bark isolation at varying times during the spring invariably resulted in abnormal secondary xylem formation (Evert & Kozlowski, 1967; Evert, Kozlowski & Davis, 1972). In the two diffuse-porous species investigated by Evert and co-workers, normal cytodifferentiation required a continual supply of assimilates from the secondary phloem. Thus it appears that normal secondary xylem differentiation requires, as a minimum, a complex combination of auxin, gibberellins, cytokinins, carbohydrates and possibly ethylene. These requirements appear to be furnished by the secondary phloem, the cambial zone itself (Sheldrake, 1973*a*) and the transpiration stream of the mature xylem. There is some evidence that the metabolic requirements for the cytodifferentiation of vessels are unlike the conditions necessary to produce fibers and tracheids (Shininger, 1970; Morey, 1973, 1974).

The initiation of a vascular cambium in roots appears to be regulated by hormones moving from the shoot system into the roots (Torrey, 1963, 1966). Although isolated roots grown in culture usually do not form secondary tissues, Torrey (1963) induced the initiation of a cambium in isolated pea roots with a combination of exogenous IAA and sucrose. Subsequently, it was shown that the optimum requirements for a similar cytodifferentiation in isolated radish roots (Loomis & Torrey, 1964) and turnip roots (Peterson, 1973) consisted of auxin, cytokinin, sucrose and *myo*-inositol.

Xylem differentiation in the interfascicular zone, i.e., the parenchymatous layer of cells between the vascular bundles, appears to represent a unique case of cytodifferentiation. These cells were observed to form tracheary elements directly, without cell division, as a result of various hormonal treatments (Siebers, 1971*b*; Fig. 3).

In concluding this review of the historical aspects of xylem differentiation it is hardly necessary to point out that this is a somewhat fragmentary outline of the subject; additional studies will be discussed in subsequent chapters. It is important to bear in mind that much of the experimental research that has been dis-

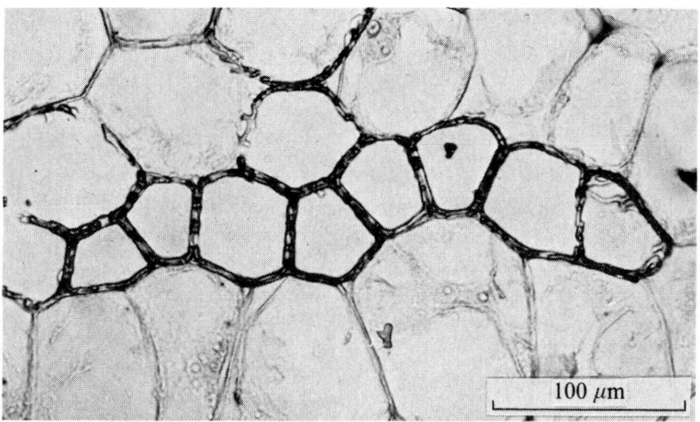

Fig. 3. Xylogenesis in an explant of interfascicular tissue isolated from the hypocotyl of a castor bean (*Ricinus communis* L.) seedling occurred directly from parenchyma cells. The explant was cultured 8 days on a basal medium supplemented with IAA (12.5 mg/1). This treatment resulted in the differentiation of the interfascicular tissue to xylem elements without an immediately preceding cell division. (From Siebers, *Acta Bot. Neerl.* **20,** 1971*b*.)

cussed concerns the chemical induction of tracheary element differentiation in parenchymatous tissues, whereas xylem differentiation in the intact plant arises from the normal development of the procambium (Steeves & Sussex, 1972). Whether or not the requirements for cytodifferentiation in both cases are identical is an open question. Tracheary elements induced to form under experimental conditions nearly always exhibit some form of reticulate or scalariform-reticulate pattern of secondary wall thickening (Fig. 1). Rarely do we find experimentally induced tracheary elements possessing helical secondary wall thickenings (Beslow & Rier, 1969; Bradley & Dahmen, 1971; Baba & Roberts, unpublished observations; Fig. 4). So far, there is no convincing evidence of any experimental technique that has resulted in the formation of xylem elements bearing annular wall thickenings. Evidently certain aspects of the cytodifferentiation of primary xylem elements have eluded us! For further information regarding experimental studies on vascular differentiation the reader is directed to Esau (1965*b*) and Steeves & Sussex (1972).

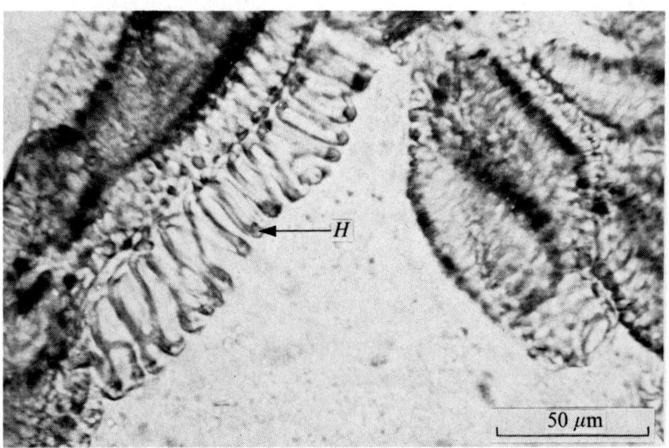

Fig. 4. Tracheary element exhibiting a helical pattern of secondary wall thickening (*H*) associated with a cluster of tracheary elements possessing the typical reticulate to scalariform-reticulate pattern. These cells were observed in an explant of lettuce pith parenchyma cultured 8 days on a xylogenic medium with constant illumination from a blue light source. A CBS 450 nm filter was employed. (From Baba & Roberts, unpublished observations.)

3

Hormones in primary xylem differentiation

Auxin, cytokinins, gibberellins, ethylene and cAMP may play diverse roles in the initiation, as well as subsequent stages, of cytodifferentiation. Our knowledge about the modes of action and interrelationships of these hormones is indeed meager. Reports of experimental results obtained with relatively high concentrations of these growth regulators should be viewed with a healthy skepticism. Although hypotheses suggesting the relative importance of transient hormone gradients, fluctuations in concentrations, feedback control mechanisms maintaining hormone levels and the hormonal regulation of sequential events are of current interest, we still lack the capability of testing most of these ideas at this time.

As far as we know, the critical variables for the initiation of cytodifferentiation are some combination of auxin and cytokinin. There is unconfirmed evidence that the cytokinin requirement is, in reality, a requirement for cAMP. Auxin may be involved in initiating the release of ethylene during the inductive period. Gibberellins can modify the pattern, extent and morphology of experimentally induced xylogenesis, but this group of hormones may be more important in the regulation of secondary xylem formation than in the ontogeny of primary tracheary elements. Gibberellins are generally ineffective in the induction of cytodifferentiation in tissue systems presumably devoid of auxin and cytokinin. Although insufficient work has been done to establish whether or not abscisic acid plays a direct role in the regulation of cytodifferentiation, certain concentrations of abscisic acid suppressed xylogenesis under in-vitro conditions (L. J. Feldman, unpublished observations).

Wareing (1971) suggested that a distinction be made between the effects of plant hormones on meristematic cells, cells about to undergo maturation and mature cells, because these three cell types differ in their respective states of determination. Meristematic cells have no prior commitments in cytodifferentiation, whereas procambial cells are predetermined, and hormones may affect the 'readout' rather than the 'programming' of the genome (Wareing, 1971). Siebers (1971*a*) demonstrated this phenomenon with small blocks of interfascicular tissue removed from the hypocotyl of castor bean seedlings. The tissue blocks were replaced in the same site at 180° inversion, and thus the tissue zone that would normally differentiate into secondary xylem was external to the future interfascicular cambium. The implantation was made prior to any evidence of mitoses in the interfascicular area. The newly formed secondary tissues were unchanged in position, i.e., secondary xylem was formed external to the interfascicular cambium, while secondary phloem was produced on the inner side. Apparently determination occurred prior to cell division in the interfascicular zone. Wareing (1971) raised the possibility that plant hormones may affect gene switching in totipotent meristematic cells, whereas in preprogrammed determined cells the effects of the hormones may depend largely on the competence of the target cells to respond. If this hypothesis is demonstrated to be true, then the hormonal response in a given cell, in regard to cytodifferentiation, would depend initially on whether or not previous gene programming had already occurred. The possibility that the point of determination is established at some time during the cell cycle will be explored in a subsequent chapter.

Although the early events in cytodifferentiation, at the molecular level, are unknown, plant hormones are capable of direct effects on RNA and protein synthesis. It is possible that they act as effectors in gene derepression, control mRNA synthesis and the functioning of mRNA, regulate the synthesis of specific tRNA fractions, facilitate the operation of RNA polymerase, influence binding of auxin to chromatin or RNA, or regulate protein synthesis at the translation stage (Davies, 1973; Key & Vanderhoef, 1973). A more comprehensive role might even be envisioned. Let us assume that there is the co-ordinated derepression at an operator site of complex *A* of the structural genes for the programmed development of a tracheary element. At a particular gene locus of complex *A* an inducer *X* might be formed which would then activate gene complex *B* for the initiation of lignification of the tracheary element. The inducers could be small protein molecules that complex with plant hormones (Matthysse & Phillips, 1969; Matthysse, 1970; Matthysse & Abrams, 1970). Such a model has been proposed for flower development (Heslop-Harrison, 1967).

At the molecular level we are concerned with the activation of a specific fraction of the genome involved in tracheary element formation. The term 'epigenotype' has been used to designate that part of the genome which, under the appropriate conditions, can be expressed with the exclusion of the functioning of other epigenotypes involved in the formation of other cell types (Abercrombie, 1967). Do auxin and cytokinin somehow activate that particular epigenotype which results eventually in the formation of a new tracheary element? What stimuli activate the epigenotypes for all other cell types found in higher plants?

Auxin

As revealed by Jacobs' early work (1952, 1954; Jacobs & Morrow, 1957), auxin undoubtedly plays a dominant role in the initiation of xylogenesis. Other factors, however, are required and these may limit the onset of cytodifferentiation. Young (1954) found that the application of auxin would not replace the role of the second youngest primordium in xylogenesis occurring below the shoot apex in *Lupinus* seedlings. Several in-vitro studies have demonstrated that both exogenous auxin and cytokinin are necessary in order to initiate cytodifferentiation, and the interaction of carbohydrates and auxin in xylogenesis has been stressed by several research groups (see Chapter 7).

The role of auxin in cytodifferentiation remains an enigma, and a detailed examination of the current hypotheses on the mode of action of auxin lies beyond the scope of this review (see Davies, 1973). Two questions immediately concern us. What is the primary and *initial* action of auxin in the induction of cytodifferentiation? After the process is started, does auxin play any additional role(s) during the various stages of the CDS? Let us first examine the requirements for the induction of the process. What precisely is the state of the cell when it is competent to respond to auxin? Is a differentiated parenchyma cell capable of a direct transformation and redifferentiation to a tracheary element? If so, does such a cell contain all the biochemical equipment, i.e., except for hormones and nutritional requirements, necessary to undertake such a task? If not, then auxin would probably be involved in DNA transcription and enzyme synthesis. Alternatively, auxin might have some direct or indirect effect on membrane physiology, ethylene biosynthesis, or the functioning of some wall-synthesizing organelle system. Several investigators have stated that tracheary element formation can occur without any visible evidence of the occurrence of a preceding cell division. This, however, can only mean that cytokinesis was not observed; the cell may have been stimulated to undergo some unobservable events in the cell cycle. Must a resting cell revert to a meristematic condition in order to be in a 'totipotent state', and is it necessary for the cell to be in a particular stage of the cell cycle in order to initiate the process? If we assume that active cell cycling is a prerequisite for auxin action, then it would appear that auxin plays some primary role either at the transcriptional or translational level. Careful cytological studies, at the light microscope level, have revealed nuclear alterations in cells destined to form tracheary elements (Innocenti & Avanzi, 1971), but these cells may be determined already at this stage, and auxin may have completed its primary function in initiating the response. These questions will be examined more carefully in Chapter 4.

It is difficult to time a process when we have no adequate method of establishing the point of determination of the CDS. Clearly auxin is capable of acting within minutes in the stimulation of growth (Evans, 1974). Fosket (1970) examined the time requirement for DNA synthesis and auxin in relation to xylogenesis in cultured *Coleus* stem segments. The minimum time for the appearance of xylem was three days, but the maximum tritiated thymidine incorporation occurred during the second day of culture (Fig. 5). Fosket's data do not support the hypothesis that auxin regulates xylogenesis by limiting DNA synthesis. The initiation of auxin-induced xylogenesis probably occurs either somewhat before or during DNA synthesis, approximately twenty-four to forty-eight hours after culture. The induction of interfascicular xylogenesis by IAA was observed

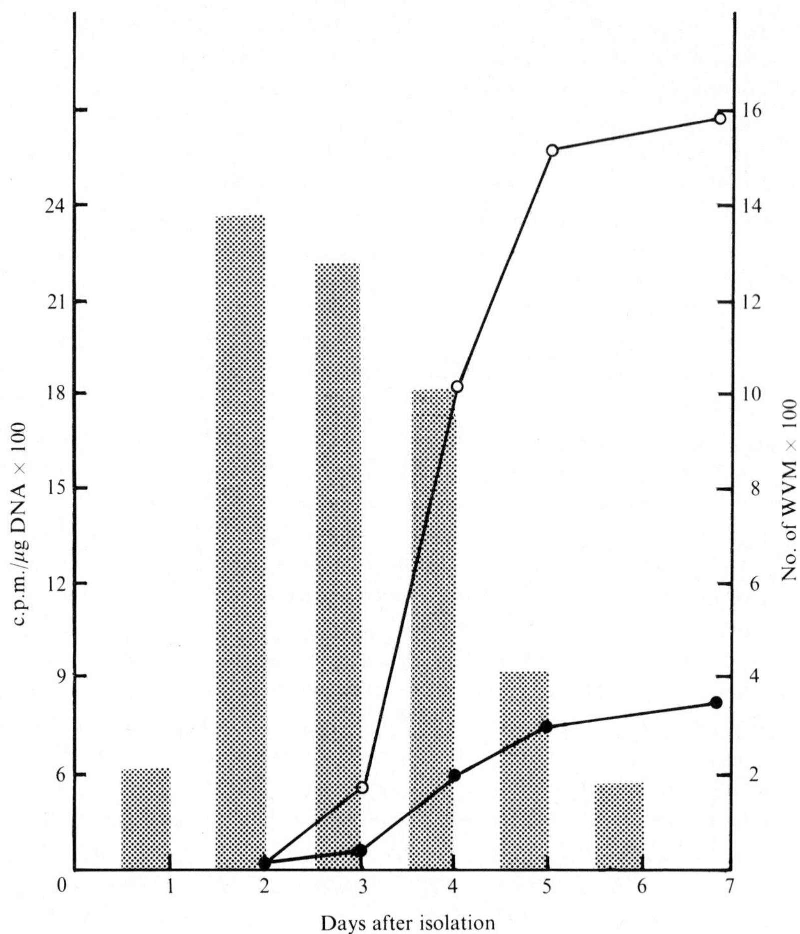

Fig. 5. The time course of DNA synthesis and xylogenesis in stem segments of *Coleus* cultured *in vitro*. Numbers of wound vessel members (WVM) were counted each day in segments cultured on a medium containing either 2 % sucrose and 1 % agar (SA) or SA medium supplemented with 0.05 mg/1 IAA (SAIAA). The mean number of WVM per segment in SA (•) and SAIAA (○) for each daily sample is given. Rates of DNA synthesis, during the same time period, were estimated from the amount of [^{3}H] thymidine incorporated into DNA during a 12-hour labeling period. The vertical bars indicate the extent of [^{3}H] thymidine incorporated into DNA in SAIAA-treated segments, the width of the bars the duration of the labeling period. (From Fosket, *Plant Physiol.* **46,** 1970.)

in bean explants in as little as eighteen hours following culture (Roberts & Baba, 1970; Fig. 6). It is important, however, to stress that timing in cytodifferentiation is relative, and depends entirely on the plant system under investigation. Most of our experimental results have come from excised tissues, because the developmental process can be precisely timed from the moment of excision. The entire

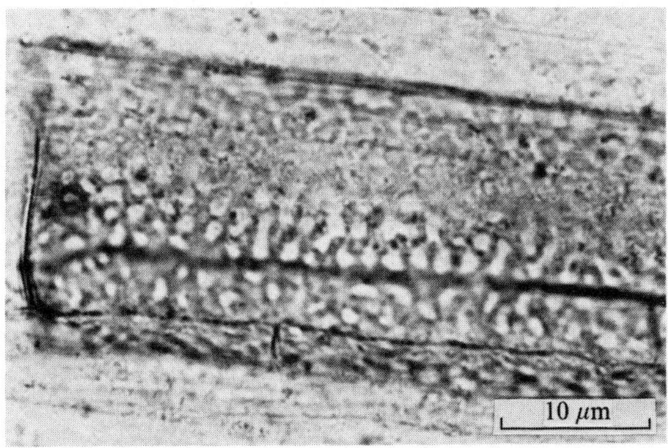

Fig. 6. Tracheary element differentiated in the interfascicular cambial zone in explant of internodal segment of *Phaseolus vulgaris* L. (cv. Oregon Giant pole bean). The explant was cultured on an agar (1 %) medium containing IAA (0.5 mg/l), kinetin (0.2 mg/l), sucrose (2 %) and colchicine (0.01 %). The colchicine was incorporated in the medium as a 'marker' for the detection of xylem differentiation. The experiment was terminated at various time intervals in order to determine the minimum period for xylogenesis in the interfascicular zone. The tracheary element in the photomicrograph was produced during an 18-hour treatment. Notice the abnormal secondary wall deposition resulting from the colchicine treatment. The cell was stained with Safranin O. (From Roberts & Baba, *Mem. Fac. Sci., Kyoto Univ., Ser. Biol.* **3,** 1970.)

CDS from cell origination to mature tracheary element can be completed in an actively growing root apex within twenty-four hours (J. G. Torrey, personal communication). Further experiments involving timing are discussed in the next section dealing with cytokinins.

Let us examine some possible roles for auxin during the various stages of cytodifferentiation. Cell enlargement is observed usually in incipient tracheary elements, and this phenomenon may be auxin-mediated, or it may be controlled by a cytokinin–gibberellin balance (Engelke, Hamzi & Skoog, 1973). An auxin–cytokinin ratio may be involved in lignification (Torrey *et al.*, 1971). Both auxin and cytokinins may be formed as degradation products in the final stages of autolysis. These hormones, thus released from dying tracheary elements, may have an autocatalytic effect on initiating cytodifferentiation in contiguous cells (Sheldrake & Northcote, 1968*a*,*b*). Certain effects of auxin on RNA and protein synthesis may be relevant. The necessity of protein synthesis for cytodifferentiation in excised internodal stem slices of *Coleus* was shown by Fosket & Miksche (1966). Actinomycin D treatment within the first forty-eight hours after explant isolation strongly inhibited xylogenesis, but treatment with actinomycin D after the first two days had no significant effect on cytodifferentiation. Although auxin may influence RNA and protein synthesis in many ways, the synthesis of enzymes involved in secondary wall metabolism during the middle stage of xylem differentiation is a definite possibility. There are conflicting reports on the significance of the effects of auxin on the regulation of β-1,4-glucan glucosyltransferase. Since cycloheximide was found largely to inhibit increased activity of the enzyme due to the presence of IAA, this suggested the necessity of protein synthesis (Hall & Ordin, 1968; Spencer, Ziola & Machlachan, 1971). Ray (1973*a*, *b*) reported that the auxin-induced increased activity of UDP-glucose:β-1,4-glucan glucosyltransferase was not prevented by either actinomycin D or cycloheximide, and he presented this as an example of a hormonal activation of a previously existing, reversibly deactivated enzyme.

The initiation and extent of cytodifferentiation is strongly influenced by auxin transport. The mobility of auxin, as reflected by xylogenic responses in parenchymatous tissue systems without pre-existing vascular tissue, will depend on at least the following factors: (*a*) the type of auxin employed, (*b*) the type of cytokinin in the medium, (*c*) the presence of a gibberellin, and (*d*) the particular tissue system under investigation. Whether or not parenchymatous explants have any inherent polarity with respect to auxin transport has not been seriously investigated, and this probably varies among plant species. The cytodifferentiation response in regard to polarity of explants of lettuce pith, and probably similar explants as well, depends partly on explant thickness (Roberts, unpublished observations). In cultured pith tissue the relative roles of diffusion and polar auxin transport are mainly unknown. Assuming that polar auxin transport occurs, does the extent of this transport vary during the development of cytodifferentiation (Sussex, Clutter & Goldsmith, 1972)? According to Sheldrake (1973*b*) the ability to transport auxin basipetally in excised stem segments of tobacco increased as secondary development occurred, whereas the ability of the pith parenchyma to transport auxin declined with age. Unfortunately, no comparative study has been made of induced cytodifferentiation in pith explants of different physiological states of maturity, i.e., from consecutively older internodes. Dalessandro & Roberts (1971) found that IAA, α-naphthaleneacetic acid (NAA) and 2,4-D, respectively, gave different patterns of xylem formation when combined with the same cytokinin in explants of lettuce pith, and that any given auxin treatment produced slightly different patterns of xylem differentiation with various cytokinins (Figs 7, 8, 9*a* and *b*). The presence of gibberellic acid had a marked effect on the pattern of auxin-induced xylogenesis, as well as modifying the response given by auxin and cytokinin (Dalessandro, 1973*b*). From the standpoint of patterns of xylem formation, the results obtained from explants of lettuce pith (Dalessandro & Roberts, 1971) and Jerusalem artichoke tuber (Dalessandro, 1973*b*) are not comparable. The possible effects of mixtures of the same type of growth regulator have not been studied. Would combinations of more than one auxin, or combinations of several different cytokinins, have additive, synergistic, or inhibitory effects on cytodifferentiation? Experimentally induced xylogenesis in the presence of pre-existing vascular tissue has been studied with whole plants and stem internodal segments. Sachs (1968, 1969) has conducted some interesting experiments on the polarity of xylem regeneration in intact pea seed-

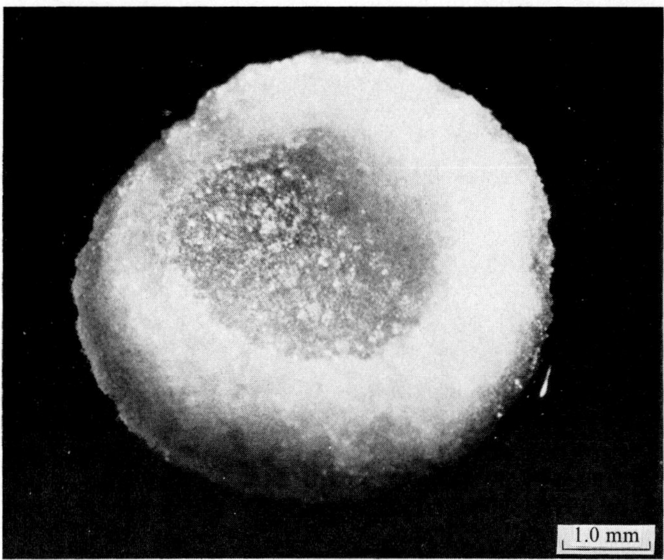

Fig. 7. Explant of pith parenchyma excised from head of Romaine lettuce (*Lactuca sativa* L. var. Romàna) after 7-days dark culture at 25 °C on a Murashige & Skoog (1962) medium containing Bacto-agar (1 %), IAA (5 mg/l), kinetin (0.1 mg/l) and sucrose (3 %). Note the extensive proliferation of callus around the periphery of the explant. (From Dalessandro & Roberts, *Am. J. Bot.* **58,** 1971.)

lings that have been surgically treated. Auxin-induced xylem strands differentiated toward pre-existing vascular tissue only if the latter tissue was devoid of an auxin source, i.e., xylogenesis was inhibited from occurring in the vicinity of vascular tissue supplied with auxin. Increased polar transport of auxin resulted from xylem regeneration, and the polarity was maintained either by the presence of auxin or auxin flow. The differentiation of strands perpendicular to the axis of this polarity was suppressed (Sachs, 1969). It is interesting to note that this phenomenon also applies to parenchymatous systems. Lateral and longitudinal strand formation were never observed together in explants of Jerusalem artichoke tuber (Dalessandro, 1973*b*) and lettuce pith (Dalessandro & Roberts, 1971), although both types of orientation were noticed (see Fig. 9*a* and *b*). The nature of this polar influence has not been studied. Isolated and centrally wounded internodes of okra (*Hibiscus esculentus* L.) and pea (*Pisum sativum* L.) were treated apically or basally with [^{14}C]IAA or [^{14}C]2, 4-D (Thompson, 1970). Apically applied auxin stimulated cytodifferentiation in the wound area; basally applied auxin did not. Approximately equivalent amounts of radioactivity were found in the middle of the wounded segments from either apical or basal applications of IAA. Unless some difference in metabolism existed between apically- and basally applied IAA, and the label applied as IAA did not remain with the auxin molecule to the same extent, it appeared that there were equivalent amounts of

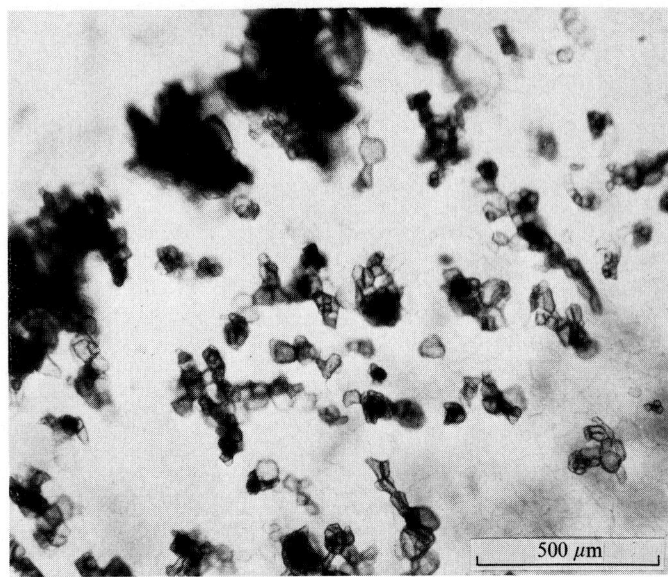

Fig. 8. Pattern of cytodifferentiation in explant of lettuce pith following treatment given in Fig. 7. A peripheral network of perpendicular strands of tracheary elements is indicated in the darkly stained areas in the upper left-hand corner of the photomicrograph. The plane of focus is in proximity to the lower surface of the explant. Numerous isolated clumps of tracheary elements are scattered throughout the explant in a zone near the surface of the agar medium. The tracheary elements were stained with Safranin O in the cleared explant. (From Dalessandro & Roberts, *Am. J. Bot.* **58,** 1971.)

IAA present for comparable levels of xylem differentiation (Thompson, 1970). Thompson (1970), however, did not investigate possible metabolic differences within the segments resulting from the two modes of application, nor did he attempt recovery and identification of the label in the cytodifferentiation region following the two treatments. These results are similar to the auxin transport experiments conducted with *Coleus,* tomato and peanut (Thompson & Jacobs, 1966; Thompson, 1968). The role of tissue polarity was examined by Forest & McCully (1971) in relation to xylogenesis in isolated segments of tobacco pith. When the morphologically apical end of the segment was fed IAA and sucrose by micropipette, vascular nodules were formed. Inverted pith explants developed vascular nodules without any apical micropipette treatment. All explants were cultured on a basal medium supplemented with NAA and coconut milk, and these xylogenic substances apparently were transported into the inverted segments and induced the formation of the vascular nodules.

Cytokinins

Similar to the physiological state of affairs with the other plant hormones, the direction of future cytodifferentiation studies with cytokinins is seriously hampered

(*a*)

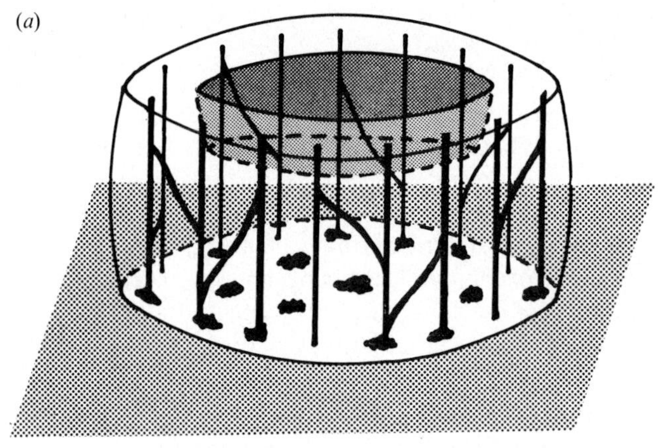

(*b*)

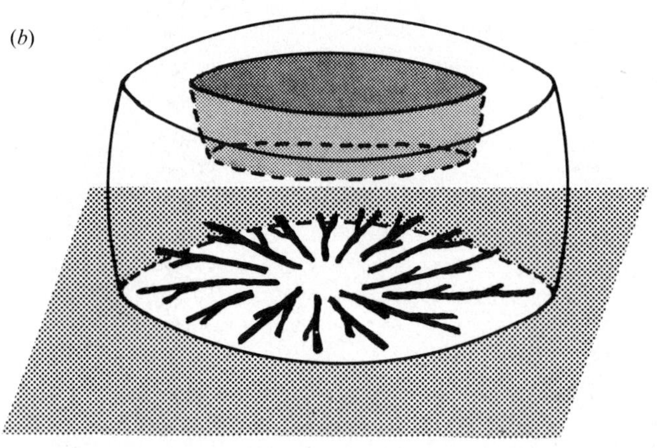

Fig. 9. Stereodiagrams showing induced xylem in explants of lettuce pith after 7-days dark culture at 25 °C on a Murashige & Skoog (1962) medium containing Bacto-agar (1 %), sucrose (3 %), auxin and cytokinin. (*a*) Differentiation of xylem occurred in a peripheral network and in basal clumps after treatment with IAA (5.0 mg/l) plus kinetin (0.1 mg/l). This pattern is shown photographically in Fig. 8. (*b*) Xylogenesis was restricted to a horizontal network of tracheary elements in explants receiving NAA (0.5 mg/l) plus kinetin (0.1 mg/l). (Adapted from Dalessandro & Roberts, *Am. J. Bot.* **58,** 1971.)

by a lack of information on the cytological site of action of the cytokinins and the underlying biochemical reactions. Indeed it is difficult to prove *in vivo* that the induction of tracheary element formation requires some cytokinin-dependent process, because most of the tissues and organs of higher plants contain appreciable levels of these substances (Torrey *et al.*, 1971).

Exogenous cytokinin was a strict requirement for the induction of cytodifferentiation in several in-vitro cultures, but the xylogenic response was only initiated in the presence of both exogenous auxin and cytokinin. This dual requirement has been demonstrated for cell suspension cultures of root callus of *Pisum* and *Convolvulus* (Torrey, 1968), stem callus of *Centaurea* (Torrey, 1968; Fig. 10), soybean callus (Torrey, 1968; Fosket & Torrey, 1969), tobacco callus (Bergmann, 1964), pea root segments (Torrey & Fosket, 1970), cortical explants of pea root (Phillips & Torrey, 1973) and pith explants of lettuce (Dalessandro & Roberts, 1971). There are some apparent exceptions to this dual requirement. Cytodifferentiation has been induced by cytokinin and sucrose, or sucrose alone, without the addition of auxin in fern gametophytes (DeMaggio, 1972), but these autonomous plants probably produce endogenous hormones. Since appreciable amounts of endogenous cytokinins have been isolated from xylem sap of *Coleus* (Banko, 1974), the requirement of only exogenous auxin and carbohydrate for the induction of tracheary elements in excised stem segments of *Coleus* (Fosket & Roberts, 1964) is not unexpected. Exogenous auxin was the only hormonal

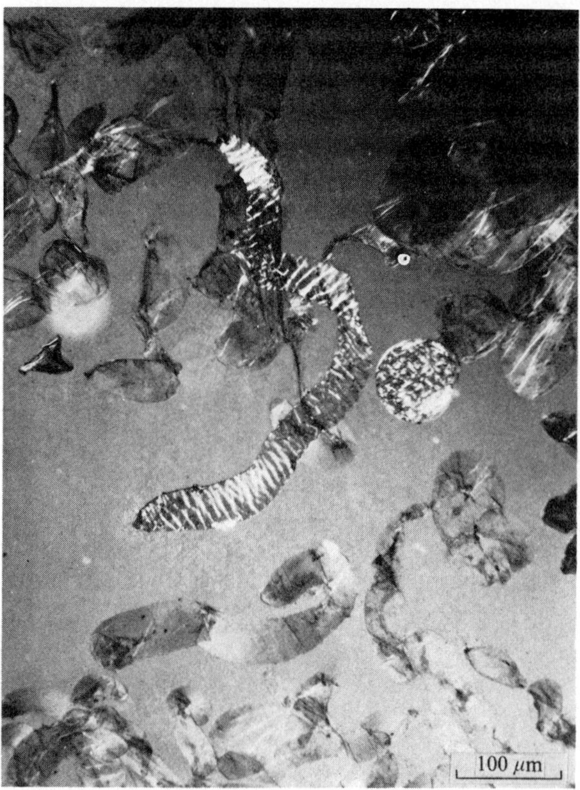

Fig. 10. A cell suspension of cornflower (*Centaurea cyanus* L.) grown 10 days on an M-6 medium following preparation from a filtered parenchyma population. Note the two tracheary elements exhibiting birefringence of the secondary wall. Numerous parenchyma cells show no evidence of secondary wall deposition. (From Torrey, *Biochemistry and physiology of plant growth substances,* 1968, p. 853. Courtesy of Runge Press, Ottawa.)

requirement for xylogenesis in explants of tubers of Jerusalem artichoke (Dalessandro, 1973*b*), but the tuber tissues contain a natural cytokinin (Nitsch & Nitsch, 1960). However, the observations that 2,4-D can mimic cytokinin activity in stimulating cell division (Witham, 1968) and cytodifferentiation (Fosket & Torrey, 1969; Dalessandro & Roberts, 1971) deserves attention. Witham (1968) has suggested three possibilities: (*a*) 2,4-D may function as a cytokinin, (*b*) it may act indirectly to increase the sensitivity of the tissue to auxin-mediated responses, or (*c*) it may stimulate the biosynthesis of cytokinins. The development of aberrant xylem elements by 2,4-D in cultured lettuce explants (Dalessandro & Roberts, 1971; Fig. 11) indicated that this growth regulator cannot perfectly match the unique role(s) of the cytokinins in this particular CDS.

The molecular requirements for xylogenesis, in connection with cytokinins, have received little attention. Regardless of the auxin employed, the effectiveness of different cytokinins in producing tracheary elements in explants of

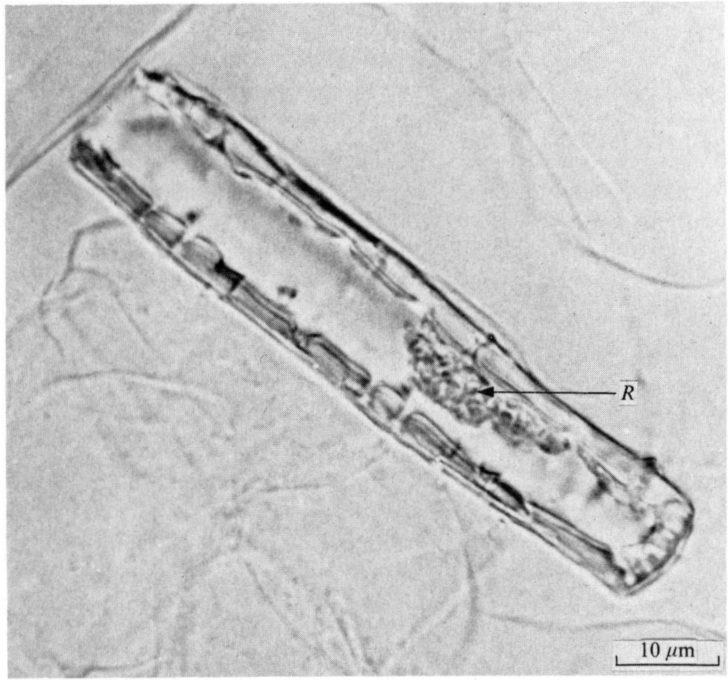

Fig. 11. Tracheary element exhibiting aberrant secondary wall deposition resulting from a 2,4-D treatment. The tracheary element was observed in an explant of lettuce pith following 14-days culture on a Murashige & Skoog (1962) medium containing 2,4-D (0.02 mg/l) in the absence of exogenous cytokinin. Apparently 2,4-D can, to some extent, mimic the cytokinin requirement in the initiation of xylem differentiation. The localized area showing a reticulate pattern of secondary wall formation (*R*) is indicated. (From Dalessandro & Roberts, *Am. J. Bot.* **58,** 1971.)

lettuce pith was in the following order: zeatin, kinetin, benzyladenine (Dalessandro & Roberts, 1971). Diphenyl urea was unable to act as a cytokinin, whereas relatively high concentrations of adenine sulfate (50 mg/l) were effective in stimulating xylogenesis in the presence of auxin in explants of lettuce pith (T. J. Banko, personal communication).

Radin & Loomis (1971) isolated an unidentified cytokinin from the xylem of developing radish roots, and this substance may have some special significance in xylogenesis. This xylem-localized cytokinin was neither a derivative of zeatin nor of isopentenyladenine, and it was inactive in the stimulation of cambial activity in cultured roots (Radin & Loomis, 1971).

Cytokinins have some regulatory function in the initiation of cytodifferentiation in addition to the stimulation of cell division. After Fosket (1968) demonstrated that cells must divide in order to initiate cytodifferentiation, the connection between these events was examined further in cytokinin-dependent cultures of soybean callus (Fosket & Torrey, 1969). Low levels of kinetin (10^{-8} M or less) stimulated cell division but tracheary elements were not formed. Kinetin concentrations of 5×10^{-8} M or above were required for the initiation of xylogenesis, and increasing concentrations of kinetin induced a progressively larger percentage of the total cell population to form tracheary elements (Fosket & Torrey, 1969; Fig. 12). Similar results were obtained by Shininger & Torrey (1974) with cortical explants of pea root. In explants of lettuce pith, IAA–zeatin and IAA–kinetin treatments produced approximately the same total cell number, but the IAA–zeatin treatment produced about seven per cent tracheary elements with respect to the total cell population while the IAA–kinetin medium was forty per cent less effective (Dalessandro, 1973*a*).

The involvement of cytokinins in cytodifferentiation was demonstrated in excised segments of pea root (Torrey & Fosket, 1970). When cultured on an auxin-containing medium in the absence of exogenous cytokinin, a callus was initiated from the diploid cells of the pericycle. Tracheary elements were never induced to form under these conditions. Segments cultured on an auxin–kinetin medium produced a callus from the cells of the cortex. Experiments with tritiated thymidine incorporation, presented at various times during the culture period, showed that the majority of the cortical cells stimulated to divide had synthesized DNA in the third day of culture prior to division, and a cytological examination of these cells indicated that they were polyploid. After five to seven days the recently divided cortical cells formed tracheary elements. A further experiment with tritiated thymidine showed that the tracheary elements were derived from the polyploid cortical cells following DNA synthesis and cell division during the kinetin treatment (see Chapter 4).

The relationships of cytokinin to the induction of polyploidy and cytodifferentiation were studied with two callus cultures (S2 and S4) of *Acer pseudoplatanus* (Wright & Northcote, 1973). The cells of S2 were mainly tetraploid and incapable of initiating cytodifferentiation, whereas the diploid cells of the S4 callus produced vascular tissues and roots. Endogenous diffusible growth factors were

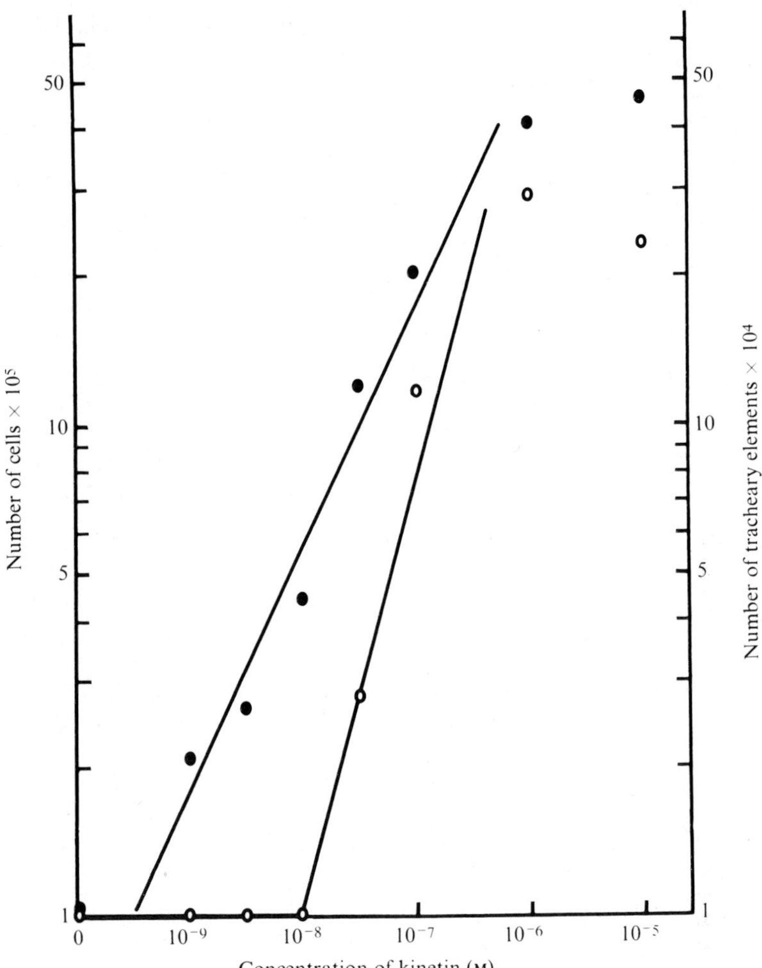

Fig. 12. The cell division and cytodifferentiation responses of an agar-grown soybean callus (*Glycine max* var. Biloxi) to various levels of kinetin. The tissue was grown on media containing 10^{-5} M NAA in the absence of cytokinin, or with kinetin concentrations varying from 10^{-9} M to 10^{-5} M. The total number of cells (•) and the number of tracheary elements formed (○) were estimated for each callus mass after 3 weeks culture on the various media. The mean cell and tracheid counts determined in the original inoculum were subtracted from the final counts obtained at the end of the 3-week period and these net values were plotted against kinetin concentration on a semi-log scale. (From Torrey, Fosket & Hepler, *Am. Sci.*, **59,** 1971; data from Fosket & Torrey, *Plant Physiol.* **44,** 1969.)

probably not involved; S2 and S4 populations cultured in the same flask produced no effect on cytodifferentiation in either callus. The cells of S2 may have either undergone mutation with an irrevocable loss of genetic capacity to differentiate, or they may have 'locked the switch' leading to gene derepression

(Torrey, 1971). The S4 callus was maintained on an NAA-containing medium in the absence of cytokinins. The cells of the S4 callus required cytokinin in order to initiate cytodifferentiation, but the cells of S4 lost the capacity to induce cytodifferentiation after two subcultures on a kinetin-containing medium. According to Fosket & Short (1973), cytokinin had quite the opposite effect in cultured soybean tissues. Although the increase in cell population was a function of the exogenous cytokinin concentration, the amount of tritiated thymidine incorporated into DNA was unrelated to the rate of cell division. The highest level of tritiated thymidine incorporation was found on the minus-cytokinin medium. Fosket & Short (1973) therefore concluded that the cells became polyploid in the absence of cytokinin, and that cytokinin was acting in the initiation of cytokinesis.

Since most studies have emphasized the synergistic effect of cytokinin and auxin in cytodifferentiation, relatively little information is available on the possible sequential requirements for growth regulators in xylogenesis. Torrey (1971) has stressed the empiricism of this approach until we have a clearer understanding of the nature of the sequential events occurring during cytodifferentiation. Attempts to relate the timing of the cytokinin requirement to xylogenesis by using cytokinin 'pulses' were made by Shininger & Torrey (1974) with cortical explants of pea root. The explants were cultured for three, six, nine, or twelve days on a kinetin-containing medium, followed by either nine, six, three, or zero days incubation in the absence of kinetin. A three-day exposure to kinetin was sufficient to induce cell division, and this event was evident by the third day. However, a minimum of six days of kinetin treatment was necessary in order to produce tracheary elements, and cytodifferentiation was not detectable until the ninth day of culture (Shininger & Torrey, 1974). Cawthon (1972) examined the sequential hormone requirements for xylogenesis in explants of lettuce pith. Explants were cultured forty-eight hours on a basal medium containing either auxin or cytokinin. Subsequently the explants were transferred for five days to a medium that contained the other plant hormone, but in the absence of the hormone that was received initially. Obviously, residual amounts of the hormone given initially will remain within the explant tissues. Explants that received auxin prior to cytokinin were completely lacking in tracheary elements. Those which received cytokinin before auxin contained tracheary elements, but the number of differentiated cells were fewer than the number formed in explants that received auxin and cytokinin simultaneously for seven days. On the other hand, the reverse sequence, i.e., auxin followed by cytokinin, favored cell division in some systems. The induction of polyploid mitoses in cultured pea root segments required auxin at some period during the first twenty-four hours of culture followed by a cytokinin requirement (Matthysse & Torrey, 1967). In cultures of tobacco callus, the optimal synchronization of mitoses was obtained when auxin was added at twenty-five hours and cytokinin at thirty to thirty-five hours after the transfer of the cells to a fresh basal medium (Jouanneau, 1971).

Rossini (1973) observed that the proliferation *in vitro* of dissociated parenchyma cells of *Calystegia* required the presence of 2,4-D prior to the addition of benzyladenine; the cells were incapable of division when the growth regulators were presented in the reverse order.

Although we have only fragmentary data, the concentration of cytokinin relative to auxin and other endogenous factors may have some significance in cytodifferentiation. Wright & Northcote (1973) examined the ratios of auxin to cytokinin effective in stimulating differentiation in a recently isolated sycamore callus (S4) composed of diploid cells. Xylogenesis and root initiation were observed with NAA–kinetin ratios ranging from 0.5 to 20; xylogenesis without root formation was mainly confined to much lower ratios, ranging from 0.025 to 0.4 (Fig. 13).

The interaction of kinetin and IAA in the orientation of wall microtubules (Shibaoka, 1974; see p. 29) may have some relationship to the synergistic effect of auxin and cytokinin on the initiation of cytodifferentiation. The kinetin inhibition of auxin-induced elongation of epicotyl sections of azuki bean apparently involved both microtubule orientation and cellulose biosynthesis (Shibaoka, 1974; Hogetsu, Shibaoka & Shimokoriyama, 1974*a,b*), and these phenomena are of prime importance to the differentiation of tracheary elements.

There is no solid evidence at this time to link cytokinins with any specific requirement for the initiation of cytodifferentiation, and the development of synchronous cultures hopefully will provide some much needed information on auxin–cytokinin interactions during cytodifferentiation. The possibility of employing anticytokinins (Hecht *et al.*, 1971) in cytodifferentiation studies should be explored.

Gibberellins

Synergism exhibited between gibberellins and other growth regulators in the stimulation of cambial activity and secondary xylem formation has been studied to a greater extent than the situation in explants containing only primary growth. Why gibberellins appear to be more effective in acting synergistically in the differentiation of secondary xylem compared with primary xylem is not clear, and may not even be relevant. The incorporation of gibberellic acid in an auxin–cytokinin medium had a greater stimulatory effect on cytodifferentiation than on cell division in cultured explants of Jerusalem artichoke (Gautheret, 1961*a*; Dalessandro, 1973*b;* Fig. 14). On the other hand, Minocha & Halperin (1974) indicated that gibberellic acid inhibited xylogenesis in explants of Jerusalem artichoke when four per cent soluble starch was provided as the carbohydrate source. Much work needs to be done on the effects of gibberellins on cytodifferentiation in isolated tissue systems. Do gibberellins *per se* play any direct role in cytodifferentiation? This is a difficult question because it is impossible to examine cytodifferentiation in any plant material involving exogenous gibberellins

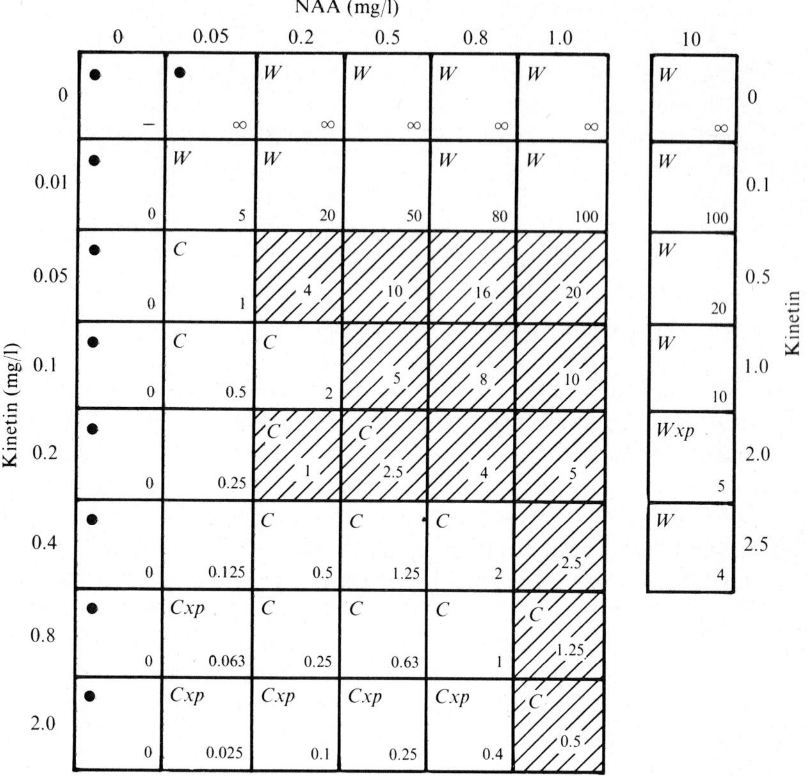

Fig. 13. The effects of various ratios of NAA and kinetin on the growth and differentiation of sycamore (*Acer pseudoplatanus* L.) S4 callus at the end of 8 weeks after subculture. The value in the lower right-hand corner of each square is the NAA : kinetin (w/w) ratio present in the culture medium. A black dot indicates death of the explant. Root initiation and the presence of vascular tissue is shown by the hatched squares. A 'white' callus was characterized by extensive growth, high friability and relatively little differentiation (*W*). Some treatments produced a 'compact' or hard callus with a red-brown pigmentation and usually some differentiated tissue (*C*). Xylem and phloem are indicated by the symbol *xp*. Roots were usually induced when the ratio was between 0.5 and 20, and the callus development could not be characterized as either *W* or *C*. At low ratios xylem and phloem cytodifferentiation were induced, and root induction was infrequent. Relatively little differentiation was found at high ratios of the hormones. (From Wright & Northcote, *J. Cell Sci.* **12,** 1973.)

with the assurance that traces of endogenous xylogenic plant hormones have been purged from the tissues. The minimum amount of gibberellin required to produce a demonstrable effect on a cell during cytodifferentiation has not been determined, but the concentration is undoubtedly very small. Some investigators have employed relatively enormous concentrations, and it is doubtful if observations from such studies are of any physiological significance. Other workers have

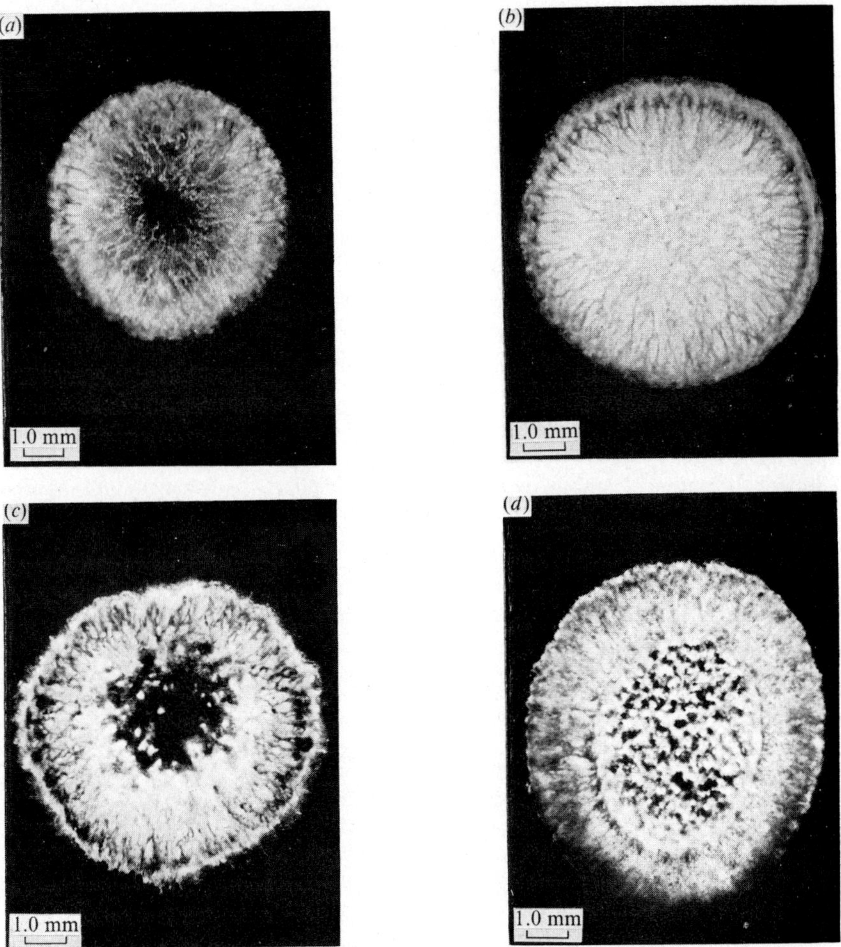

Fig. 14. Patterns of tracheary element cytodifferentiation induced in explants of Jerusalem artichoke tuber (*Helianthus tuberosus* L.) dark-cultured at 27 °C on Murashige & Skoog's (1962) medium containing Bacto-agar (1 %), sucrose (3 %) and various combinations of hormones. Tracheary elements were stained with Safranin O. (From Dalessandro, *Plant Cell Physiol*. **14,** 1973*b*.)

(*a*). Explant treated with IAA (5 mg/l). The resulting cytodifferentiation did not occur in the absence of cytokinin, since this tissue contains an endogenous cytokinin (Nitsch & Nitsch, 1960).

(*b*). Explant treated with IAA (5 mg/l) and zeatin (0.1 mg/l). Note the enhanced growth and altered pattern of cytodifferentiation compared with (*a*).

(*c*). Explant treated with IAA (5 mg/l) plus gibberellic acid (1.0 mg/l).

(*d*). Explant treated with a combination of IAA (5 mg/l), gibberellic acid (1.0 mg/l) and zeatin (0.1 mg/l). The percentage of the total cell population that differentiated as tracheary elements for each of the four treatments was as follows: (*a*), 13.54; (*b*), 16.55; (*c*), 15.84, and (*d*), 19.43.

autoclaved gibberellin-containing media, but steam sterilization can degrade more than ninety per cent of the biological activity of a gibberellin solution (Bragt & Pierik, 1971). Wound-induced gibberellin production (Rappaport & Sachs, 1967) may modify the results obtained with tissue explants, and an attempt to suppress gibberellin biosynthesis with antigibberellins was only partially successful (Siebers & Ladage, 1973). If gibberellins play any direct function in cytodifferentiation, it is not likely to be a limiting one. As far as we can determine, gibberellins are incapable of exhibiting a xylogenic response in the absence of other plant hormones.

Gibberellin blocked the formation of bud-producing meristematic nodules in tobacco cultures, although xylogenesis and lignification remained unaffected by its presence (Murashige, 1964). The requirements for the initiation of organ primordia and vascular nodules are unrelated (Halperin, 1969; T. A. Thorpe, personal communication). Doley & Leyton (1970) observed that gibberellic acid applied to wound callus on *Fraxinus* stems did not increase growth or tracheary element formation. Cambial explants of *Salix* grown on an NAA-containing medium produced a homogeneous parenchymatous callus (Saussay, 1969). Although small vascular nodules were observed with the addition of kinetin, the combination of gibberellic acid and NAA produced two highly organized zones of peripheral vascular bundles containing xylem, phloem and cambium. The formation of giant thin-walled pitted tracheids was observed in pith parenchyma explants of lettuce (Cawthon, 1972; Fig. 15) and tobacco (Snijman, 1972)

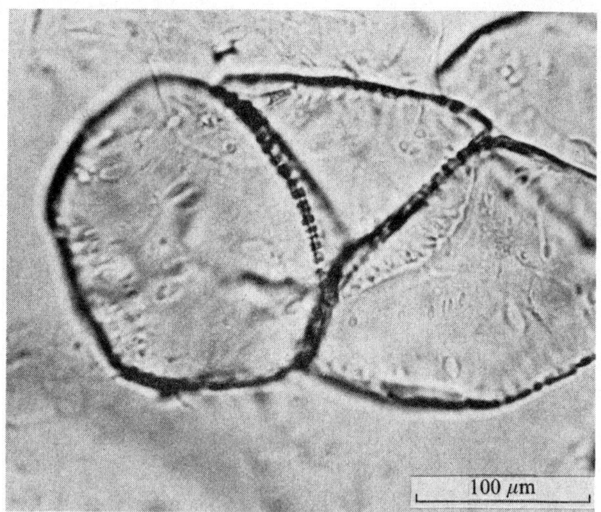

Fig. 15. Cytodifferentiation induced in the presence of gibberellic acid in explant of lettuce pith parenchyma. Explant was dark-cultured 8 days on a Murashige & Skoog (1962) medium containing IAA (5 mg/l), kinetin (0.1 mg/l), gibberellic acid (1.0 mg/l), sucrose (3 %) and Bacto-agar (1 %). These relatively large balloon-shaped cells were characterized by pitted walls and weak staining with Safranin O. The explants also contained tracheary elements with reticulate and scalariform-reticulate wall thickening.

treated with gibberellic acid. These unique cells were observed in small numbers only in the presence of both gibberellic acid and auxin, and these cultures also contained the typical tracheary elements with scalariform-reticulate secondary wall thickenings. Gibberellin, acting in some manner with auxin, apparently delays cytodifferentiation and facilitates cell expansion. The relatively uniform wall expansion resulting in these balloon-shaped cells may arise from a dual effect of auxin and gibberellin promoting both lateral and longitudinal wall growth simultaneously. Torrey *et al.* (1971) suggested that gibberellins may influence the determination of the secondary wall pattern, and recent evidence suggests that this involves the interaction of gibberellin and microtubules. The gibberellin-induced elongation of lettuce hypocotyl was inhibited by colchicine (Sawhney & Srivastava, 1974). The colchicine-induced inhibition of coleoptile elongation was partially reversed by gibberellin, and a competitive interaction at the microtubule level between colchicine and gibberellin was suggested (Fragata, 1970). The concept of gibberellin reacting, in some manner, with microtubules is supported by the findings of Shibaoka (1972). In epicotyl sections of azuki bean (*Azukia angularis*) gibberellin enhanced IAA-induced growth, but had no stimulatory effect without exogenous auxin. Specifically, gibberellin stimulated longitudinal elongation, but cell expansion in a direction transverse to the axis of the organ was suppressed. The application of colchicine completely reversed the effect of gibberellin without affecting the IAA-induced lateral expansion of the sections. Since the direction of cell expansion was controlled by the orientation of microfibrils (Green, 1969), gibberellin may promote stem elongation by acting on microtubules. In epicotyl segments of azuki bean treated with gibberellin plus IAA, the wall microtubules were arranged in a position transverse to the cell axis. Wall microtubules in segments treated with kinetin plus IAA were parallel to the cell axis, whereas wall microtubules in segments treated with IAA alone were randomly oriented (Shibaoka, 1974). The relationship of these findings to observations on hormonal synergism in cytodifferentiation is difficult to interpretate, since the combination of IAA, cytokinin *and* gibberellin produced the optimum xylogenic response in explants of Jerusalem artichoke (Dalessandro, 1973*b*). The auxin-induced elongation of epicotyl sections of azuki bean was stimulated by gibberellic acid and inhibited by kinetin. Both the gibberellin synergism and the kinetin antagonism were reversed, without affecting auxin-induced elongation, in the presence of inhibitors of cellulose biosynthesis (Hogetsu *et al.*, 1974*a*,*b*). The possible effects of gibberellins and cytokinins on the initiation of cellulose biosynthesis during auxin-induced cytodifferentiation would be of considerable interest. The endoplasmic reticulum may play specific roles during xylem differentiation (see Chapter 6), and the synthesis of the endoplasmic reticulum in barley aleurone cells was demonstrated to be regulated by gibberellin and abscisic acid (Evins & Varner, 1971; Vigil & Ruddat, 1973).

In tobacco plantlets, gibberellins promoted the elongation of developing vascular tissue of the leaf, whereas cytokinins tended to inhibit the elongation of the same tissue (Engelke *et al.*, 1973). Since the greater emphasis has been placed on the role of auxin in comparison with the other hormones, perhaps we have overlooked the possibility of a gibberellin–cytokinin ratio as playing a basic role in the regulation of tracheary element elongation during the normal ontogeny of the developing vascular system.

An indirect effect of gibberellin on xylogenesis involves the induction of hydrolyase activity resulting in the formation of sugar monomers necessary for wall synthesis. Studies unrelated to cytodifferentiation indicate that this area should be explored. Invertase activity in excised *Avena* stem was increased with gibberellin, and the substrate end-products of the enzymatic degradation probably played significant roles in the regulation of invertase levels in the tissues (Kaufman *et al.*, 1973). In dwarf-pea internodes gibberellin stimulated amylase, β-fructofuranosidase and starch phosphorylase (Broughton & McComb, 1971). The injection of glucose and glucose derivatives into pea internodes mimicked the effects of gibberellins on cell elongation, cell division and cell wall synthesis, and it was considered that the overall effect of gibberellin was to provide more substrate for cell metabolism and wall synthesis (Broughton & McComb, 1971). Can the observable effects of gibberellins on xylem differentiation be interpreted in terms of a carbohydrate requirement for wall metabolism? Probably not, since no one has succeeded in mimicking the effects of gibberellins on cytodifferentiation by employing any combination of sucrose or glucose and fructose, i.e., substrate and end-products of invertase activity.

Waller & Burström (1969) employed an auxin bioassay involving the promotion of cambial activity in decapitated pea seedlings. Gibberellic acid was also effective in stimulating cambial activity in this bioassay. The addition of delcosine, a diterpenoid of the lycoctonine-type, inhibited approximately twenty-two per cent of the gibberellin-induced xylem differentiation response. However, the addition of ajaconine, a diterpenoid of the atisine-type skeleton, had no effect on the gibberellin response in the bioassay. The authors suggested that the inhibitory effect of delcosine might be due to either a competition with gibberellin for active sites or to feedback control of gibberellin biosynthesis. Further work with delcosine might provide some information on the role of gibberellin in xylem differentiation.

Tannins might be employed to study gibberellin-induced xylem differentiation. Tannins are known to act as gibberellin antagonists, leaving auxin activity unaffected (Corcoran, Geissman & Phinney, 1972). Durzan, Chafe & Lopushanski (1973) observed that spruce callus, cultured under conditions of alternating light and dark (coupled with high and low temperatures), produced high levels of tannins compared with cultures maintained under constant environmental conditions. Xylogenesis occurred in the cultures maintained under the constant conditions, but tracheary elements were completely absent from the cultures kept under the alternating environmental conditions. High tannin formation was associated with low starch formation, and this may have been a factor in the inhibition of cytodifferentiation (Chafe & Durzan, 1973). The inhibition of cell division, however, may have suppressed xylogenesis, since Jacquoit (1947) found that cambial cells would not proliferate in media containing more than traces of tannins. Earlier, Gautheret (1961*b*) reported that callus of Jerusalem artichoke remained undifferentiated at low temperatures, and xylogenesis did not occur below 17 °C. Whether or not these cultures grown at low temperatures contained tannins was not indicated.

In short, there is no clear-cut evidence at this time that gibberellins play an inductive role in cytodifferentiation. Practically no critical research has been done on these hormones in the area of primary vascular differentiation. Aside from gibberellic acid, few physiologists have performed any experiments on cytodifferentiation with any of the other gibberellins. Since this group of hormones affects the morphology of differentiating tracheary elements, it is surprising that a comparative study of the effects of various gibberellins on cytodifferentiation has not been made. Are gibberellins in the xylem sap, possibly products released from autolyzing tracheary elements in the final stages of cytodifferentiation, actively functional in cytodifferentiation (Sheldrake, 1973*a*)? Or do they act indirectly to 'modify' cytodifferentiation as they pass through the vascular tissues from a site of production to a site of utilization? Because of a preoccupation with the biochemistry of the cereal aleurone, we have sadly neglected critical studies in some interesting areas of developmental physiology. In the review by Jones (1973) on the physiological roles of gibberellins, there is no discussion of any of the recent research on the effects of this group of hormones on vascular differentiation.

Ethylene

Physiologists have been reluctant to grant hormone status to the physiological action of ethylene (see discussion by Abeles, 1972). This conservative attitude has arisen because, as a gas, ethylene is not readily transportable from an organ of synthesis to a site of action. The necessity for transportation in order to qualify as a plant hormone is unwarranted. The impossibility of demonstrating that auxin, cytokinins and gibberellins do not have *any* hormonal functions within the cells in which they are produced is obvious. On the other hand, it is definitely possible that ethylene is released by certain cells and tissues with hormonal functions in adjacent cells and tissues, assuming that diffusion is considered an acceptable vehicle of transportation of a hormone. It seems that the mode of transportation is completely irrelevant, but that the interaction of ethylene with other plant hormones at extremely low concentrations in the regulation of plant growth and development is of significance.

There are several experimental studies which have indicated that ethylene is capable of regulating cytodifferentiation and that it may be of fundamental importance in tracheary element differentiation. Unfortunately, most experimental studies have employed concentrations of ethylene far exceeding physiological levels. The observation that fifty parts per million of ethylene suppressed xylogenesis (Apelbaum, Fisher & Burg, 1972) is in the realm of toxicology rather than physiology. The production of ethylene by the secondary xylem of *Pinus radiata* exhibited seasonal fluctuations, and ethylene in this tissue may be involved in phenol synthesis and heartwood formation (Shain & Hillis, 1973). Ethylene gas (5 to 20 ppm) applied to the trunks of *Pinus radiata* and *Liquidambar styraciflua* trees produced more secondary xylem when treated for ninety-nine days or longer than did untreated controls, whereas no increase was observed in trees treated for shorter periods of time (Neel, 1970). The application of ethylene (1.1 ppm) almost completely eliminated cambial activity in cultured radish roots, but lower concentrations were not tested (Radin & Loomis, 1969). Exogenous IAA (10^{-5} M) and benzyladenine (5×10^{-6} M), respectively, increased the rate of endogenous ethylene production, and a synergistic effect on ethylene formation was observed when both auxin and cytokinin were applied simultaneously. This synergism may involve the suppression of the conversion of IAA to IAA-conjugates, with the higher level of free IAA responsible for the enhanced production of ethylene (Lau & Yang, 1973). The latter effect on the conjugation of auxin was mediated by kinetin; it is unknown whether benzyladenine and endogenous cytokinins function in this manner. The induction of vascular cambia in cultured radish roots by the application of exogenous auxin and cytokinin may be a result of endogenous ethylene production by the excised organ (Loomis & Torrey, 1964). The diageotropica (*dgt*) single gene mutant of tomato required exogenous ethylene for normal growth and development, and ethylene concentrations as low as 5 nl/l completely normalized the growth characteristics of the *dgt* mutant (Zobel, 1973). Secondary vascular differentiation was disturbed in the *dgt* mutant, whereas primary vascular differentiation appeared normal. The secondary xylem of the mutant completely lacked large vessels (Zobel, 1974; Figs 16, 17).

Experimental studies on cytodifferentiation with excised plant tissues probably involve influences by wound-induced ethylene production (Abeles & Abeles, 1972), and trauma-induced ethylene biosynthesis has been reported in secondary xylem tissue (Cooper, 1972). The application of mechanical stress to branches of white pine, apple, and peach trees increased the ethylene in the internal atmosphere of the wood by over fifty per cent (Leopold, 1972). Leaf epinasty following clinostat treatment was associated with increased ethylene production (Lyon, 1972; Leather, Forrence & Abeles, 1972). Enhanced xylem differentiation in wounded *Coleus* shoots subjected to clinostat treatment doubtless was related to stress-induced ethylene production (Roberts & Fosket, 1962).

Exogenous ethylene induced parallel increases in both the hydroxyproline content and peroxidase activity in the covalently bound cell wall protein of pea shoots. Therefore it was speculated that ethylene increased the cytoplasmic hydroxylation of proline that led to higher levels of hydroxyproline-rich peroxidases of the cell wall (Ridge & Osborne, 1970, 1971). The enhanced level of hydroxyproline possibly leads to an increased cross-linking by glucosidic bonds

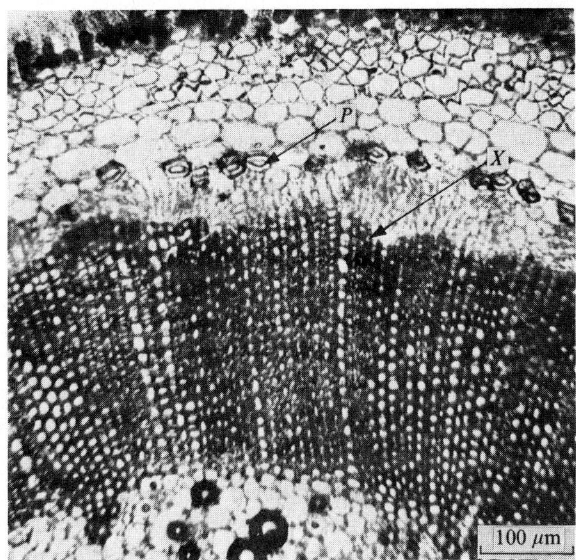

Fig. 16. Transverse section of the stem of the diageotropica mutant (*dgt*) of tomato (*Lycopersicon esculentum* Mill.). This mutant requires exogenous ethylene for normal growth and development. The vascular tissues of the mutant are characterized by abnormally thick phloem fibers (*P*) and the lack of large vessels in the secondary xylem (*X*). (Courtesy of R. Zobel. Reproduced by permission of the National Research Council of Canada from *Can. J. Bot.* **52,** No. 4, 1974, pp. 735–41.)

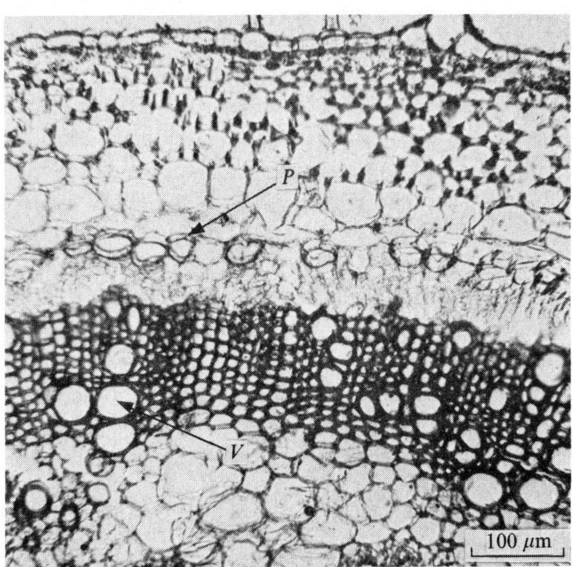

Fig. 17. Transverse section of the stem of a normal tomato (*Lycopersicon esculentum* Mill.), the isogenic parent (VFN8) of the *dgt* mutant shown in Fig. 16. Compare the development of phloem fibers (*P*) and large vessels (*V*) with the stem section shown in Fig. 16. (Courtesy of R. Zobel. Reproduced by permission of the National Research Council of Canada from *Can. J. Bot.* **52,** No. 4, 1974, pp. 735–41.)

of hydroxyproline-rich proteins with polysaccharides in the cell wall, and this has implications for wall plasticity and cell growth (Osborne, Ridge & Sargent, 1972). Exogenous proline may stimulate xylogenesis in cultured explants of *Coleus* under certain experimental conditions (Roberts & Baba, 1968*b*).

The timing of the action of ethylene in a physiological process is difficult to determine. Although ethylene and auxin have completely different physiological effects on cell wall deposition (Sargent, Atack & Osborne, 1973), these differential effects appear to be sequentially related (Sargent, Atack & Osborne, 1974). Sargent *et al.* (1974) have proposed that in the control of growth in epidermal cells of *Pisum*, the initial response is an IAA-induced cell expansion followed approximately twelve hours later by an ethylene-induced cell wall thickening accompanied by enhanced peroxidase activity. At some period during the first twelve hours auxin doubtless initiates ethylene biosynthesis (Abeles, 1972). The most effective time for the application of exogenous ethylene in order to induce stem abscission in explants of *Phaseolus vulgaris* was after twenty-four hours from the time of excision. The minimum time for cell division in explants of Jerusalem artichoke was approximately thirty hours (Yeoman & Davidson, 1971) to thirty-six hours (Dalessandro, 1973*b*) from the time of excision. We may postulate that the peak effectiveness of ethylene in cytodifferentiation is either during or shortly after the first peak of mitotic activity. The first tracheary elements have been observed in explants of Jerusalem artichoke at approximately seventy-two hours (Dalessandro, 1973*b*).

Linkins, Lewis & Palmer (1973) examined the relative effects of auxin and ethylene on cytodifferentiation in explants of *Phaseolus*. Indoleacetic acid (10^{-2} M in lanolin paste) stimulated ethylene production in the explants, and cell division and cytodifferentiation occurred. After exhausting the vessels of ethylene to less than 0.02 nl/ml, extensive cell division and cytodifferentiation were still evident, but we can assume that the auxin-induced endogenous ethylene produced within the tissue was sufficient for the induction of cytodifferentiation. Treatment of bean petioles with 10^{-5} M IAA is the highest auxin concentration which can be employed without stimulating ethylene production over that observed in control petioles (Linkins *et al.*, 1973). This level of auxin produced some pit formation in parenchyma cells, but cell division and cytodifferentiation responses characteristic of the 10^{-2} M auxin treatment were not observed.

Inherent problems are associated with experimental design in studying the possible effects of ethylene levels on cytodifferentiation. If tissues are cultured in a closed system, the production of endogenous ethylene shortly reaches toxic levels. At concentrations above 100 nl/l ethylene is highly effective in blocking xylogenesis in cultured explants of lettuce pith (R. W. Zobel, unpublished observations). There are two main sources of endogenous ethylene in explants exhibiting cytodifferentiation: an initial rise associated with wounding, and an auxin-induced biosynthesis. Because of this endogenous biosynthesis, it is exceedingly difficult to examine the possible effects of low levels of ethylene on cytodifferentiation. The optimal atmospheric concentration of ethylene for the growth and development of tomato plants is approximately 10 nl/l with a threshold at 0.7 nl/l (Zobel, 1974). Xylogenesis in lettuce pith explants gives a similar optimum level (R. W. Zobel, unpublished observations), and further investigations on the interrelationship of ethylene and cytodifferentiation should employ concentrations of this magnitude. The inhibition of ethylene production in cultured tissues by the use of benzyl isothiocyanate (Patil & Tang, 1974) may prove useful in future studies. The unusual effects of carbon dioxide on cytodifferentiation reported by Bradley & Dahmen (1971) may involve ethylene antagonism (see Chapter 8). The physiological effects of relatively large dosages of ethylene in higher plants was recently reviewed by Abeles (1973).

Adenosine 3′,5′-cyclic monophosphate (cAMP)

There is a report, as yet unconfirmed, that the role(s) of cytokinin in the cytodifferentiation of tracheary elements may be replaced by cAMP (Basile *et al.*, 1973). These workers presented evidence to suggest that the induction of xylogenesis in explants of lettuce pith could be initiated in the absence of exogenous cytokinin by adding either cytokinesin I (Wood & Braun, 1967; Wood, Lin & Braun, 1972) or 8-bromoadenosine 3′,5′-cyclic monophosphate (8-Br-cAMP). Previously Wood *et al.* (1972) showed that cytokinesin I inhibited cAMP phosphodiesterase activity. Presumably in the Basile *et al.* (1973) experiment on xylogenesis, the cytokinesin I functioned by maintaining a relatively high level of endogenous cAMP. Cell division in tobacco pith tissue was stimulated by 8-Br-cAMP, and this synthetic analogue was resistant to native cAMP phosphodiesterase activity (Wood & Braun, 1973). Because cAMP is degraded readily by endogenous cAMP phosphodiesterases, an attempt was made to block the activity of this enzyme by adding the phosphodiesterase inhibitor theophylline (Wood *et al.*, 1972) to the xylogenic medium along with cAMP. A weak xylogenic response was observed in the presence of cAMP, theophylline, and auxin in the absence of cytokinin (Basile *et al.*, 1973). In the latter experiment cAMP was employed at a concentration of 100 μM. It is quite possible that cAMP was metabolized to adenine, and that the xylogenic response was an adenine-induced response. At a concentration of approximately 100 μM, adenine sulfate can function as a cytokinin and produce a moderate-to-weak xylogenic response in the presence of auxin in explants of lettuce pith (T. J. Banko, personal communication). Ferré (1971) reported that adenine sulfate stimulated xylogenesis in callus cultures of *Parthenocissus tricuspidata*. Basile *et al.* (1973) did not indicate if a similar experiment had been conducted with either adenine sulfate or adenylic acid substituting for cAMP. Kessler (1973) reported that cAMP induced xylogenesis in soybean callus cultures grown on Miller's (1961) medium in the presence of IAA and cytokinin. It seems improbable that cAMP has provided any unique response in this system, because Fosket & Torrey

(1969) described similar results with soybean callus cultures grown on Miller's (1961) medium in the presence of kinetin and auxin (NAA or 2,4-D) without any exogenous cyclic mononucleotides. In short, there is no substantial evidence at this time that cAMP has a hormonal function in the initiation of xylem differentiation beyond a rather weak adenine-type response.

Abscisic acid

Abscisic acid strongly inhibited the differentiation of cambial derivatives in *Phaseolus* seedlings when applied alone or in the presence of IAA, gibberellic acid, or both IAA and gibberellic acid (Hess & Sachs, 1972). A similar effect was observed in explants of Jerusalem artichoke tuber cultured on a xylogenic medium containing NAA and benzyladenine (Minocha & Halperin, 1974). Some preliminary experiments by L. J. Feldman (personal communication) have revealed that xylogenesis in cortical explants of pea root was inhibited by certain concentrations of exogenous abscisic acid. These observed effects of abscisic acid on cytodifferentiation may involve the blockage of nucleic acid metabolism (Bex, 1972; Stewart & Smith, 1972). Since abscisic acid exerts a regulatory function in several physiological processes involving auxin–cytokinin interactions, further cytodifferentiation studies should definitely be undertaken with abscisic acid.

The *most probable* hormonal critical and controlling variables during the four ontogenetic stages of tracheary element cytodifferentiation have been summarized in Table 1.

TABLE 1. *Hormonal controls during the ontogenetic stages of tracheary element differentiation* (modified after Torrey, 1953).

Stage of cytodifferentiation	Possible hormonal or chemical regulation
Competence of the target cell	
The cell has a transient competence to commence cytodifferentiation following DNA synthesis and cell division. The point of determination probably involves a period of the cell cycle. Protein synthesis is required during the first 48 hours.	Auxin and cytokinin play roles in the cell cycle and in gene expression, and cytokinin has some hormonal function in addition to that of initiating cell division. The replacement of the cytokinin requirement by cAMP has been suggested, and this finding requires further study. Ethylene probably plays some role following the induction by auxin. Gibberellic acid exhibits a synergism in some auxin–cytokinin function(s), and gibberellin may be a controlling rather than a critical variable. To date there is no substantial evidence on either sequential control or on hormonal ratios for cytodifferentiation.

Stage of cytodifferentiation	Possible hormonal or chemical regulation
Cell enlargement	
The cell exhibits high variability in the extent and regulation of expansion. The deposition of primary wall materials during expansion requires the synthesis of primary wall monomers, i.e., nucleotide diphosphate sugars of the D-galactose series. The assembly and positioning of microtubules for secondary wall patterning occurs. Protein synthesis occurs. DNA replication involving endoreduplication and gene amplification may occur.	Cell enlargement may involve auxin or possibly a gibberellin–cytokinin balance. The activation of β-1,4-glucan glucosyltransferase by auxin occurs. Auxin and cytokinin may be transported to adjacent cells for an autocatalytic initiation of cytodifferentiation. There is evidence *pro* and *con* concerning the role of cytokinin in endoploidy. The microtubules may be regulated in some manner by gibberellin in establishing the pattern of secondary wall deposition, and gibberellin probably induces hydrolyase activity (e.g., amylase, invertase) which releases sugar monomers for wall synthesis. Ethylene may influence wall thickening and stimulate an increase of hydroxyproline-rich peroxidases in the developing cell wall.
Secondary wall formation and lignification	
Cell enlargement and the synthesis of secondary wall monomers (nucleotide diphosphate sugars of the D-glucose series) occur concomitantly with diminished epimerase activity. Microtubules play a role in the orientation, but not in the synthesis or deposition of cellulose microfibrils. The synthesis of cellulose and hemicellulose occurs. Lignin metabolism, involving the activation of enzymes associated with lignin synthesis and the synthesis of lignin precursor pools, occurs prior to the deposition of the polymer. The controls are set for future cell autolysis and hydrolyase activation.	Hydrolyase activity induced by gibberellic acid may provide sugar monomers for wall metabolism. Auxin may regulate hemicellulose production. Auxin, cytokinins, gibberellins and ethylene may have diverse roles in the regulation of lignification. Sucrose, operating as a substrate rather than a hormone, may act as a controlling variable for monosaccharide nucleotide formation.
Autolysis of cell contents and selective dissolution of wall areas	
Degradative enzymes are possibly released by lysosomes. The breakdown of cytoplasmic membranes occurs and organelle functions cease. The degraded protoplast is eventually removed from the cell lumen. An enzymatic degradation of selected wall areas (nonlignified?) occurs, and contiguous cells may provide some physiological regulation in the final stage of maturation.	There is no evidence for hormonal regulation of this stage. The autolysis of the protoplasmic contents may release auxin and other xylogenic hormones for the initiation of cytodifferentiation in contiguous cells.

4

Role of the cell cycle

Nuclei exhibit a cyclic behavior in dividing cells. Two events in the nuclear cycle are mitosis and the synthesis of DNA in preparation for the following mitosis. These events are separated by gaps of time in order for the nucleus to prepare for either mitosis or DNA synthesis. The period of time just prior to mitosis (M) is referred to as G_2, and the corresponding time interval before DNA synthesis (S) is termed G_1 (Howard & Pelc, 1953). After nuclear division and cytoplasmic division have been achieved, the two daughter cells, starting in G_1, then repeat the process. The term 'cell cycle' represents the sequence between one of these events and a similar event in the next cell generation, and the complete cycle consists sequentially of G_1, S, G_2 and M (Fig. 18). Wimber (1963) described techniques that may be employed for the measurement of the duration of these different periods.

A close relationship exists between cell division activity and the differentiation of tracheary elements (Fosket, 1968, 1970, 1972; Fosket & Torrey, 1969), but the nature of this relationship remains obscure. In this chapter we will investigate evidence concerning the possibility that the cell cycle plays some role in the initiation of cytodifferentiation, and, more specifically, whether or not the point of determination is restricted to some particular period of the cell cycle. It is not clear whether a cell becomes committed to a particular pathway of cytodifferentiation during the meristematic activity that produced it, or if it is determined at some subsequent rest period.

Holtzer (1970) has formulated some concepts based on research in his laboratory with myogenic, chondrogenic and erthropoietic cells. Genetic reprogramming, which permits dividing cells to produce progeny with different synthetic capabilities, is coupled to a sequence of cell cycles. Cell cycles which yield daughter cells with the same genetic program as the parent cells have been termed 'proliferative' cell cycles, whereas cell cyles producing daughter cells having different genetic programs have been termed 'quantal' cell cycles. An asymmetrical cell division could account for daughter cells with different phenotypes (Stebbins & Shah, 1960; Pickett-Heaps & Northcote, 1966*b*), and this is an example of a quantal cell cycle. Another example involves the 'decision' of the cell to start the synthesis of myosin during G_1 instead of engaging in the preparatory reactions leading to DNA synthesis. In this case, the event responsible for categorizing the cell cycle as proliferative or quantal occurred after the actual

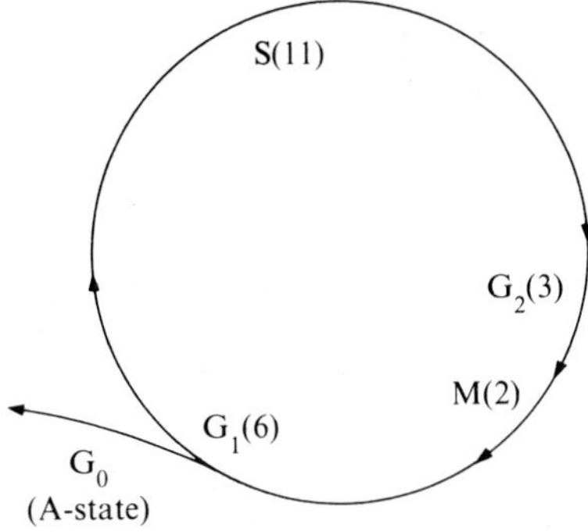

Fig. 18. Diagram of the successive periods of the cell cycle. Following DNA synthesis (S) the cell enters G_2 and mitosis (M) occurs. After nuclear and cytoplasmic division the daughter cells enter G_1. Some evidence suggests that a nonproliferative condition known as G_0 may exist. The term 'cell cycle' refers to the sequence between one of these periods and a similar period in the next cell generation. In the above diagram the length of each arc of the circle, as indicated by the position of the arrows, is approximately proportional to the duration of the phase of the cycle. The value for each period is given in hours. For example, *Tradescantia* whole root tips (21°C), measured with [^{3}H] thymidine pulse labeling, have given G_1 (6), S (11), G_2 (3) and M (2). (The example cited is from Wimber, *Am. J. Bot.* **53,** 1966.)

division. Holtzer (1970), in emphasizing a neglected feature of cytodifferentiation, states: 'Basic changes in nuclear activity are initiated by cytoplasmic cues. The stepwise differences in the cytoplasms of cells in an evolving lineage stem from a modest number of derepressions which are coupled in an obligatory fashion to particular quantal cell cycles. Genes define, permit and sustain biosynthetic programs; they do not initiate them.'

The concept of a cell 'at rest' in a nonproliferating condition needs some explanation. The hypothesis that a cell may rest in a G_0 period (Lajtha, 1963, 1967; Quastler, 1963), which is neither G_1 nor G_2 in its metabolic characteristics, has stimulated considerable debate. Some research workers have argued that G_0 represents a special biochemical condition in which processes occur that have no counterpart when the cell is in the cell cycle, whereas others have claimed that the resting cell is blocked at some point in G_1 and thus G_0 is not a unique period (Smith & Martin, 1973). The concept of G_0 is not unlike the A-state hypothesis of Smith & Martin (1973). These workers have proposed that some time after mitosis all cells enter a state (A) in which their metabolism is not directed toward replication, and that a given cell may remain in the A-state for an indeterminate length of time. On leaving A-state the cells enter B-phase, i.e., resume cell cycling, and their subsequent metabolic activities are deterministic and directed toward DNA replication (Smith & Martin, 1973). Cooper (1971) examined biochemical changes accompanying the resumption of growth in regenerating mammalian liver, phytohemagglutinin-stimulated lymphocytes, and estrogen-stimulated rat uterus. In all of these systems there was an early initiation of RNA and protein synthesis, but a lag of from one to two days before the beginning of DNA synthesis. Since this lag in DNA synthesis was much longer than the typical G_1 period of these cells, Cooper (1971) considered it improbable that the resting cell was simply suspended in some early stage of G_1. Farber & Baserga (1969) have presented more convincing evidence that G_0 is a *bona fide* period. Hydroxyurea has been employed as a specific DNA synthesis inhibitor (Young & Hodas, 1964; Kihlman, Erikkson & Odmark, 1966; Young, Schochetman & Karnofsky, 1967). Farber & Baserga (1969) found that putative G_0 cells, e.g., hepatocytes and cells of the kidney tubules, recovered from hydroxyurea treatment within a few hours and began DNA synthesis. On the other hand, actively proliferating cells, e.g., epithelial cells of the crypts of the small intestine, cells of the germinal centers of lymph nodes and of the basal layer of the epidermis, never recovered from hydroxyurea treatment but eventually died (Farber & Baserga, 1969). The authors suggested the existence of a metabolic diversity between G_0 and G_1, and that the S period may have some significant differences in various types of tissues. In any case, the latter results should be confirmed with FUdR, since Haber & Schwarz (1972) have seriously questioned the effectiveness and specificity of hydroxyurea as a blocking agent for DNA synthesis.

Van't Hof (1974) stated that cytodifferentiation in the cells of root meristems was accompanied by cell arrest in a nonrandom manner, but that cell arrest was preferential, favoring 4C in some species, e.g., *Vicia* and *Pisum,* and 2C in other species, e.g., *Helianthus* and *Triticum*. If meristematic cells of a given species favor arrest in G_1, the cells will preferentially arrest at the 2C level during maturation. On the other hand, carbohydrate-starved cells of the meristematic tissues generally arrest in G_2, and the mature cells of these populations will consist mainly of 4C DNA levels (Van't Hof, 1974). The problem of deciding whether a given cell is arrested in either G_1 or G_2 may be resolved by the selection of a species in which there are structural differences between these two periods that can be visually identified. Nuclear differences characteristic of G_1 and G_2 were found in *Allium carinatum* (Nagl, 1968*a,b,* 1970) and in *Allium flavum* (Nagl, 1970).

If we assume that some period of the cell cycle is involved in the initiation of cytodifferentiation, what event might occur that would signal the beginning of the process? The signal would probably be the appearance *de novo* of a particular enzyme, and there is current interest in mapping alterations in enzyme activity throughout the cell cycle in synchronous systems (Mitchison, 1971; Yeoman & Aitchison, 1973). If increased enzyme activity is a reflection of increased enzyme protein, then we are concerned about the synthesis of one or more specific proteins during the cell cycle. These proteins might be involved with cell wall metabolism, or they could be involved in gene expression, e.g., operating at the derepression level (Matthysse, 1970; Matthysse & Abrams, 1970). Another possibility is that certain endogenous hormones are synthesized or released during the cell cycle. It is unknown whether a differentiating cell produces its own hormonal needs or whether the hormones are imported, and we have no evidence at this time to link any particular period of the cell cycle with hormone production.

Returning to our assumption that the cell cycle is involved in cytodifferentiation, can we associate any particular period with the beginning of the process? This is a difficult question with many ramifications. If the cell cycle plays only an *indirect* role, then a definitive answer simply does not exist. There are indications that DNA synthesis precedes cytodifferentiation (Stockdale & Topper, 1966; Fosket, 1970); during the period when both DNA and histone synthesis occur, the genes for cytodifferentiation could be freed from their existing repressors and become available for programming (Ebert & Sussex, 1970). Mitchison & Creanor (1969) found that sucrase, alkaline phosphatase and acid phosphatase showed a doubling in the rate of enzyme synthesis during G_2 in the cell cycle of *Schizosaccharomyces*. They indicated that there was a time delay between chemical replication of the genome and 'functional replication', i.e., the appearance of functionally new protein. This suggests that gene programming may occur during S, and it also raises the question of whether we should consider the point of determination at the level of gene programming with the production of mRNA or with the appearance of the newly synthesized and functional protein. Gene programming refers to mRNA production for a specific protein, and thus the cells must 'differ' from any nonprogrammed cells at this time. In Mitchison & Creanor's (1969) experiment, the point of determination and the initiation of cytodifferentiation should be considered to start with gene programming, which probably occurred during the S period. In certain micro-organisms the S period may lead to cytodifferentiation. Dawes, Kay & Mandelstam (1971) found that sporulation in *Bacillus subtilis* could only be induced if the cells were starved when DNA was replicating; otherwise the cells were committed to another round of vegetative growth. Holtzer & Bischoff (1970) considered the possibility that gene derepression for myosin synthesis occurred during DNA synthesis in the presumptive myoblast, but that initiating factors for transcription or translation were not formed until the following G_1.

The subtle distinctions between determination and differentiation remain a debatable issue among zoologists (Gross, 1968; Cook, 1974). For example, the imaginal disk cells of the larvae of *Drosophila* are predestined to produce certain adult tissues. This determined condition occurs early in the development of the larvae and is carried for many cell generations before metamorphosis leads to the overt expression of differentiation with the production of proteins characteristic of the mature tissues (Ephrussi, 1972). Normally-differentiated animal cells, in many cases, can be propagated *in vitro* without the loss of differentiated functions characteristic of the parent tissue (see review by Green & Todaro, 1967). Ephrussi (1972) has indicated that the manipulation of the culture medium can often result in the disappearance of overt signs of differentiation but that the same differentiation can be re-expressed by the re-establishment of the appropriate conditions. Consequently there is an argument for the concept of the inheritance of a determined rather than a differentiated state (Abercrombie, 1967; Brown & Dawid, 1969; Ephrussi, 1972). Gross (1968), however, is of the opinion that no distinction should be made between determination and differentiation, i.e., commitment to a course of change is essentially inseparable from the molecular mechanisms involved in the realization of that particular change.

In cortical explants of pea root the cortical cells exhibited endoreduplication and continued DNA synthesis without entering the M period; cytodifferentiation evidently occurred sometime between successive DNA cycles (Phillips & Torrey, 1973). Further experiments by Phillips & Torrey (1974) involved the determination of the DNA content of differentiating tracheary elements in cortical pea root explants (see following section on endoreduplication). Tracheary elements with 2C DNA must represent diploids that ceased cycling in G_1. The data, however, do not prove that cytodifferentiation occurred in G_1 for the higher C values. For example, 4C may represent diploids in G_2 or tetraploids in G_1; 8C may be tetraploids in G_2 or octoploids in G_1. Endoreduplication of DNA takes place outside the normal operation of the cell cycle, and the terminology of G_1 and G_2 seems inappropriate. The phenomenon of endoreduplication will be discussed in a separate section at the end of this chapter.

The experiments of Van't Hof & Kovacs (1972) suggest a possible technique for studying the relative roles of G_1 and G_2 in cytodifferentiation. If the meristems of cultured pea roots are temporarily starved of carbohydrates, the cells are arrested in G_1 and G_2 (Van't Hof, 1966, 1967; Webster & Van't Hof, 1970). Starved meristems which had been arrested were termed 'stationary phase' meristems (Van't Hof, 1968*a*). Since the starved cells stopped preferentially in G_1 or G_2, these periods apparently gave some quality not found in either S or M. Van't Hof (1971) proposed that the cell cycle is regulated by two 'principal control points'; one in G_1 that regulates the G_1-to-S transition and another in G_2 that controls the G_2-to-M change. A comparison was made of the distribution of cells in the various periods of the cell cycle of root meristems of *Pisum* and *Vicia* after seventy-two hours carbohydrate starvation (Van't Hof & Kovacs, 1972). In *Pisum,* both before and after starvation, approximately the same proportion of the cells were either in G_1 or G_2. In *Vicia*, under similar conditions, fewer cells were in G_1 and considerably more cells were in G_2 compared with the carbohydrate control (Fig. 19). A series of experiments on cytodifferentiation, based on the latter finding, might be of considerable interest. These could follow the form outlined below:

(*a*) A preliminary experiment to establish whether or not *Vicia* cortical cells are capable of the induction of tracheary element formation similar to the cortical explant system of *Pisum* (Phillips & Torrey, 1973).

(*b*) A second experiment to determine whether or not the cortical cells of *Vicia*, after induction to division by an auxin–cytokinin treatment, could be

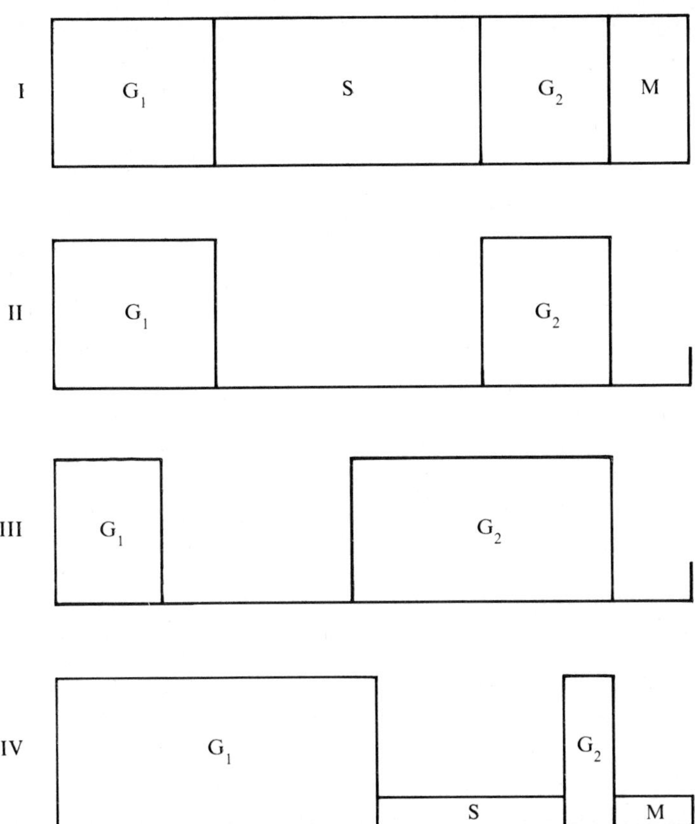

Fig. 19. Diagrams representing the cell distribution in each of the periods of the cell cycle in primary root meristems of *Pisum, Vicia* and *Helianthus* before starvation (I); in *Pisum* (II) and *Vicia* (III) after 72 hours of starvation; and in *Helianthus* after 48 hours of starvation (IV). Note that after starvation of *Pisum* (II) and *Vicia* (III) no cells were in S or M. Starvation of *Vicia* (III), unlike *Pisum* (II), served to increase the proportion of the cells arrested in G_2 compared with G_1. In the case of *Helianthus* (IV) cells were present in all four periods, although the majority were arrested in G_1. (From Van't Hof & Kovacs, *The dynamics of meristem cell populations,* 1972.)

brought to a stationary phase condition by carbohydrate starvation. The rationale for this experiment is based on the assumption that the cortical cells 10 mm from the root tip (*Pisum* technique of Phillips & Torrey, 1973) will be arrested in relatively the same cell cycle periods as the meristematic cells examined by Van't Hof & Kovacs (1972), and this point would have to be established.

(*c*) A third experiment to involve a comparison of the relative numbers of tracheary elements formed in cortical explants of *Vicia* after carbohydrate starvation with those in continuously fed explants. If higher numbers of tracheary elements were found in the explants that were temporarily starved, this would suggest that cytodifferentiation in this system was more closely associated with G_2 than G_1.

Evans & Van't Hof (1973) described a substance promoting cell cycle arrest in G_2 in root meristems; the substance was produced in the cotyledons and transported to the root system. The pea cotyledon factor promoted cell arrest in G_2 in shoots and roots of *Pisum* and in roots of *Vicia*, but was unable to promote cell arrest in G_2 in the meristem root cells of *Helianthus* and *Triticum*, which normally show predominant arrest in G_1 (Evans & Van't Hof, 1973). Substances produced in mammals by the cycling cells that induce cell arrest in G_1 or G_2, termed chalones, have been described (Frankfurt, 1971; Bullough, 1972; Rytömaa, 1973).

The complex relationships between enzyme synthesis, during the cell cycle, and differentiation, have been briefly reviewed by Mitchison (1973). Although a substantial number of the enzymes investigated showed an apparent continuous synthesis throughout the cell cycle, most of them were periodically synthesized in some particular part of the cell cycle. If the enzyme was stable, the synthesis pattern was stepwise; if the enzyme was unstable, the pattern peaked at the time of synthesis, followed by a drop as the enzyme was degraded (Mitchison, 1969, 1971). Step and peak enzymes associated with DNA synthesis generally start their periods of synthesis near the beginning of S. However, the synthesis period of enzymes other than those concerned with DNA synthesis were found throughout the cell cycle. In fact, the synthesis patterns of step enzymes were found to continue when DNA synthesis had been inhibited in bacteria, yeast, and mammalian cells; apparently there was *no* connection between DNA replication and the regulation of periodic enzyme synthesis (Mitchison, 1973). One suggested mode of control involves 'oscillatory repression', or end-product repression by negative feedback (see review by Donachie & Masters, 1969). For example, a substance accumulates during its synthesis and eventually will repress the synthesis of one of its own biosynthetic enzymes. This theory predicts that there should be continuous synthesis rather than steps if the enzyme is fully repressed or fully derepressed (Mitchison, 1973), and this is open to question. Why the oscillations in synthesis should have the same frequency as the periods of the cell cycle is probably the most difficult feature of this idea to explain.

Another theory for the control of enzyme synthesis postulates that 'linear reading' or 'sequential transcription' occurs, i.e., that the genes are transcribed during the cycle in the same order as their linear sequence on the chromosomes (Halvorson *et al.*, 1966; Halvorson, Carter & Tauro, 1971). The genes would be transcribed for a brief period only during each cycle by an RNA polymerase moving along the chromosome. Although several reports support this theory, it is difficult to reconcile linear reading with the inducibility of enzymes. Linear reading predicts that enzymes should be capable of induction only for a particular period of the cycle, but evidence suggests that enzymes can be induced at all stages of the cell cycle in bacteria, fission yeast and *Chlorella* (Mitchison, 1973).

Although conclusive evidence is lacking, numerous investigators have observed that certain parenchymatous cells appeared to exhibit a direct transformation or redifferentiation to a tracheary element without an immediately preceding cell division (D'Amato, 1953; Sachs, 1969; Dalessandro & Roberts, 1971; Gee, 1972; Basile *et al.*, 1973). In the nondividing cell there may be DNA synthesis and gene programming for cytodifferentiation. In some cases, a cell may be programmed for cytodifferentiation during an earlier meristematic period, and then function in a predifferentiated state (Runner, 1970) while awaiting hormonal read-out of the genome for tracheary element differentiation. This may be the case for interfascicular cells (Siebers, 1971*a,b;* see Chapter 5) and for protoxylem differentiation in gamma plantlets (Foard & Haber, 1961; Haber & Foard, 1964). It is important to realize that the establishment of determination for a given CDS does not necessarily mean that there must be an immediate and complete expression of that determination. The transmission of a 'determined-but-unexpressed state' through repeated cell divisions has been reported for both plants and animals (Heslop–Harrison, 1967; Ursprung, 1969; Wareing, 1971), although how a preselected gene program is transmitted through repeated mitotic divisions is not understood. The view that an immediately preceding mitosis should not be considered as a necessary prerequisite for the expression of cytodifferentiation is clearly illustrated by the experiments of Foard (1960). The ontogeny of foliar sclereids of *Camellia japonica* has been described by Foster (1944). These sclerenchymatous idioblasts originated from parenchyma cells of the leaf during the late stages of leaf maturation after all cell division activity had ceased. The first detectable changes that occur in the parenchymatous sclereid initials are increased nuclear and nucleolar sizes. Subsequently, there is a period of localized primary wall growth in the form of tubular branches, and this is followed by the formation of a massive secondary wall (Foster, 1944). Foard (1960) demonstrated that the cytodifferentiation of these sclereids could be experimentally induced by leaf wounding. Parenchyma cells that normally would have differentiated to photosynthesizing mesophyll cells were induced to form foliar sclereids in the absence of cell division. Various aspects of cytodifferentiation and organogenesis in the absence of cell division have been reviewed by Foard (1970).

The apparent incompatibility between cell proliferation and cytodifferentiation has stimulated considerable debate among animal physiologists. Cameron & Jeter (1971) have examined this problem with the various developing systems in embryonic chick tissues. In some cell systems, e.g., skeletal muscle and primary motor neurons, cell proliferation and cytodifferentiation were clearly incompatible. However, in some systems, e.g., cardiac muscle, primary series of erythrocytes, corneal stroma, liver, chondrocytes, pancreas and lens, cells with differentiated characteristics were also capable of cell proliferation. Cameron & Jeter (1971) stated that in those cases where cytodifferentiation was compatible with DNA synthesis and mitosis, the evidence indicated that only after DNA synthesis and mitosis had ceased did relatively large amounts of specialized metabolic products appear. This, however, does not answer the question of whether or not specialized protein synthesis and DNA synthesis occur simultaneously. Apparently DNA synthesis and collagen synthesis were not antagonistic in collagen-synthesizing fibroblasts (Davies, Priest & Priest, 1968; Cameron & Jeter, 1971). Differentiating muscle cells produced myosin, actin and meromyosin but did not synthesize DNA, as shown by a lack of incorporation of tritiated thymidine into DNA; muscle cells engaged in DNA synthesis did not synthesize myosin as indicated by the lack of binding of fluorescent myosin antibody (Stockdale & Holtzer, 1961). In the developing pancreas, however, there are mitotic cells which contain zymogen (Wessells, 1964), and DNA synthesis has been reported in antibody-producing cells (Mäkelä & Nossal, 1962). The synthesis of DNA, as shown by the incorporation of tritiated thymidine into nuclei of mammalian cardiocytes, was not inhibited by the presence of contractile proteins (Goldstein, Claycomb & Schwartz, 1974). The meristematic cell, similar to an embryonic animal cell, is a highly specialized and uniquely differentiated system (Malamud, 1971). These cells spin through the cell cycle with the efficient production of numerous proteins necessary for cell division: thymidine kinase, DNA polymerase, proteins of the mitotic apparatus and histones. Malamud (1971) has suggested that the apparent mutual exclusiveness of cell division and the synthesis of specialized proteins may be due to limitations on substrate and energy supplies, and that the shift in metabolism to the production of specialized protein synthesis diminishes the supply available for the synthesis of cell division needs.

The relation between cell expansion and cytodifferentiation is of considerable interest. Barlow (1971) has suggested that intercellular communication may play a dominant role in xylogenesis. The cytodifferentiation of a lineage of xylem elements in the root apex could be maintained by three factors: the plasmodesmatal interconnections across transverse walls, the synthesis of cell enlargement factors, and the rapid growth rate which prevents mitosis. Differentiating xylem cells in the root apex have a more rapid growth rate than adjacent cells in the meristem (Barlow, 1969*b*). Barlow (1971) has considered that these differentiating xylem cells are incapable of influencing their lateral neighboring cells, since the number of lateral plasmodesmatal connections is diminished by the rapid growth rate. This suggests that the pattern of development is regulated by the more basal xylem cells in the same lineage, and that the latter cells provide the necessary stimuli for xylogenesis in the uncommitted cells at the distal end of the lineage (Barlow, 1971). Different combinations and concentrations of auxin, cytokinin, gibberellin and carbohydrate are capable of inducing a wide variety of sizes and shapes of tracheary elements under in-vitro conditions. The differentiated cells may be extremely small, balloon-shaped, or rectangular and elongated. Evidently the extent and direction of cell expansion, involving primary wall metabolism, is highly variable. Perhaps the onset of secondary wall deposition, which would prevent any further enlargement, can be postponed in some cells. If certain combinations of plant hormones can alter the timing of the cell cycle, and if the production of certain secondary-wall-metabolizing enzymes depends on certain stages of the cell cycle, the variation in the extent of cell enlargement might be explained. A bigger or longer tracheary element has simply had a longer period of time to expand. There are several possibilities concerning cell enlargement and cell cycling for a parenchymatous cell in an explant cultured on a xylogenic medium. The cultured cell may undergo cell cycling with virtually no cell expansion, cytodifferentiation appearing at some period during cell cycling. Certain 2,4-D treatments induced the formation of extremely minute tracheary elements (Roberts, unpublished observations), in which cytodifferentiation must have occurred immediately after cell division and probably occurred in G_1. The second possibility is that the cultured cell exhibits cell cycling and some manner of cell enlargement, with cytodifferentiation occurring at some time during the process. Cells enlarge in meristematic tissues after cytokinesis while in G_1, S and G_2, but there is no evidence that cells continue to enlarge during the relatively short metaphase period (J. G. Torrey, personal communication). In micro-organisms, wall material may be incorporated throughout the cycle with the exception of the M period (see discussion by Mitchison, 1971). Maksymowych & Kettrick (1970) observed that cell enlargement reached a peak at the termination of DNA synthesis and cell division during leaf development; further DNA synthesis was not required for cytodifferentiation. Specific inhibitors for DNA synthesis have been demonstrated to prevent auxin-induced cell expansion in excised tobacco pith tissues (Maheshwari & Noodén, 1971). On the other hand, the enlargement and maturation of normal tracheary elements was observed in wheat seedlings in which mitotic activity had been completely suppressed with gamma irradiation (Foard & Haber, 1961; Haber & Foard, 1964). The issue concerning DNA synthesis and cell enlargement is debatable, and the arguments are tangential to the present review.

The experimental evidence relating hormones to a particular period in the cell cycle is meager at this time. Banerjee (1968) studied DNA synthesis in cells of the root meristem of *Zea mays* dwarf mutants (d_1 and d_5). These plants, dwarf when homozygous, grew to normal size following treatment with gibberellic acid. Pulse labeling revealed that the duration of the cell cycle in both mutants was lengthened by approximately twenty-five per cent compared with the normal-sized heterozygotes. Both the G_2 and S periods were longer than normal in the mutants. Whether or not gibberellin treatment had any effect(s) on the length of the cell cycle was not reported by Banerjee (1968). Interphase in *Allium* was shortened with combinations of exogenous auxin and kinetin (González-Fernández *et al.*, 1972) and with kinetin alone (Guttman, 1956). In *Pisum* root meristems, Van't Hof (1968*b*) found that auxin plus kinetin impaired the entry of cells into S and their development during the S period, and that kinetin increased the duration of G_2. Tobacco cell suspensions were brought to division synchrony by a temporary lack of cytokinin followed by an addition of cytokinin to the medium after a suitable lag period. These experiments suggested that a single event in the cell cycle may be controlled by cytokinins (Jouanneau, 1971; Péaud-Lenoël & Jouanneau, 1971). Jouanneau & Tandeau de Marsac (1973) observed that cytokinin added to a suspension culture of tobacco induced synchronous divisions after an eighteen-hour lag period; cytokinin was not essential to mitoses and cytokinin-dependent events were completed before mitoses occurred. The evidence also indicated that the cytokinin-dependent events were independent of DNA synthesis because DNA synthesis proceeded normally in the presence of auxin, with or without cytokinin, for at least the time required for one completion of the cell cycle (Jouanneau & Tandeau de Marsac, 1973). Exogenous cytokinin was necessary for the initiation of cytodifferentiation in cortical cells in pea root explants; the process was apparently initiated some time after DNA synthesis had occurred (Torrey & Fosket, 1970).

The timing of cytodifferentiation, in relation to DNA synthesis, was examined in excised stem segments of *Coleus* (Fosket, 1970). Tracheary elements were first observed after three days of culture, and this three-day lag period was the most active time for DNA synthesis, as indicated by tritiated thymidine labeling (see Fig. 5). Xylogenesis was also inhibited concomitant with the inhibition of DNA synthesis by specifically blocking the activity of thymidylate synthetase with 5-fluorodeoxyuridine (FUdR; Matthysse & Torrey, 1967). Exogenous thymidine completely reversed the effect of FUdR (Matthysse & Torrey, 1967; Fosket, 1970). The inhibitor had no effect on xylogenesis when added to the medium after the third day of culture, i.e., after the peak period of DNA synthesis had passed. Inhibitor studies, however, are open to interpretation, because FUdR blocks organelle DNA synthesis as well as nuclear DNA synthesis, and certain processes may be more dependent on mitochondrial DNA synthesis than on nuclear DNA synthesis (Degani & Atsmon, 1970); Degani, Atsmon & Halevy, 1970). Although exogenous auxin greatly stimulated xylogenesis, it had no effect on the time period of tritiated thymidine incorporation, and auxin only slightly enhanced tritiated thymidine incorporation into DNA. Apparently auxin does not regulate cytodifferentiation by limiting DNA synthesis. Because colchicine was as effective as FUdR in blocking xylogenesis during the early part of the culture period, Fosket (1972) suggested that the completion of the cell cycle was necessary for the cytodifferentiation process.

In tobacco pith, cell division was found invariably to precede cytodifferentiation (Sussex *et al.*, 1972). Fosket & Short (1973) examined the role of cytokinin in a cytokinin-requiring strain of cultured soybean tissue. The amount of tritiated thymidine incorporation into nuclear DNA was *not* related to the rate of cell division, and cultures lacking cytokinin incorporated more tritiated thymidine than actively dividing cultures in the presence of the hormone. The authors suggested that the increased nuclear DNA synthesis was associated with endoreduplication in the absence of cytokinin, and that cytokinin was involved as a specific trigger for mitotic division (Fosket & Short, 1973).

The remarkable xylogenic response of explants of Jerusalem artichoke tuber to the combined presence of auxin, cytokinin and gibberellin (Dalessandro, 1973*b*) may be directly related to the fact that at least the first two cell cycles are partially synchronized (Yeoman & Evans, 1967).

Nagl (1972) demonstrated that the cell cycle in root meristems of several species of *Allium* was selectively arrested in various periods by the application of different combinations of growth regulators. This information may be of some value in future studies, but it gives us no insight into the physiological roles of endogenous hormones during the normal operation of the cell cycle.

Endoreduplication

During the early stages of cytodifferentiation of tracheary elements, progressive doubling in the amount of nuclear DNA has been reported (Lorz, 1947; D'Amato, 1952*b*; List, 1963; Phillips & Torrey, 1973). This DNA doubling reflects the endoreduplication of chromosomes, and the phenomenon has been termed somatic polyploidy, polysomaty, and endopolyploidy. An increased number of chromatids per chromosome, termed polyteny, has also been observed in differentiating tracheary elements following hormonal treatment (Nagl & Rücker, 1972; see review by Pearson, 1974). Several workers have expressed the view that chromosomal endoreduplication *per se* is not causally related to the initiation of cytodifferentiation (D'Amato, 1965; Partanen, 1965; Cutter & Feldman, 1970), although chromosomal variability may limit or affect the capability of cultured tissues to express organogenesis (Torrey, 1967). Root cortical tissues in leguminous species displayed differences with regard to endoreduplication according to species and even varieties (Libbenga & Torrey, 1973). Endoreduplication was observed in the root meristem cells of several plants without any indication of incipient cytodifferentiation (D'Amato & Avanzi, 1948; D'Amato, 1952*a*; Tschermak-Woess, 1960). On the other hand, cytodifferentiation occurs in haploid systems (Tulecke, 1967; DeMaggio, 1972) and in plants which normally contain no polyploid cells (Partanen, 1965). Experiments on the induction of chromosome doubling by endoreduplication, prior to mitosis, in cortical explants of pea have suggested that auxin and cytokinin may provide the necessary stimulus to initiate the process (Libbenga & Torrey, 1973). Endoreduplication and cytodifferentiation were examined in the cortical cells of pea root explants (Phillips & Torrey, 1973) with a tissue punch technique (Libbenga & Torrey, 1973). In the presence of auxin and kinetin, the nuclei of the cortical cells showed tritiated thymidine incorporation beginning between twenty-four and thirty-two hours; mitoses commenced at approximately forty-eight hours. Initially all mitoses were tetraploid, but as the experiment progressed, the proportion of tetraploid cells decreased and an octaploid population increased. Tracheary elements were differentiated initially on the seventh day and continued to exhibit differentiation up to three weeks in culture. Data from the work of Phillips & Torrey (1973) were subjected to analysis by a theoretical model proposed by Meins (1974), and the experimental and calculated values for the numbers of tracheary elements were in close agreement. Cytodifferentiation evidently occurred sometime between successive DNA cycles in the cortical cells. Subsequently, Phillips & Torrey (1974) examined the DNA content of Schiff-stained nuclei located within the differentiating tracheary elements of the cortical cells. The DNA content, as measured with a Vickers M-85 scanning microdensitometer, varied considerably. The relative amounts of DNA per nucleus were given as 'C values'. A normal diploid cell after DNA synthesis, and before the next mitosis, has the 4C amount of DNA (Swift, 1950). Phillips & Torrey (1974) found that the percentages of tracheary elements containing different C values were as follows: 2C (3 per cent), 4C (57 per cent), 8C (39 per cent) and 16C (1 per cent). Although most of the differentiating tracheary elements were either tetraploid or octaploid, cytodifferentiation evidently occurred in approximately three per cent of the diploid cells. Cutter & Feldman (1970), with regard to the formation of trichoblasts of *Hydrocharis,* also expressed the view that the initiation of cytodifferentiation was not associated with any particular DNA level. Wright & Northcote (1973) examined two different strains of cultured sycamore callus. The predominantly diploid strain exhibited cytodifferentiation when cultured on an auxin–cytokinin medium, whereas the predominantly tetraploid strain failed to undergo cytodifferentiation. The diploid strain required exogenous cytokinin in order to initiate cytodifferentiation, but after two transfers on a cytokinin-containing medium, lost the capability of exhibiting differentiation. The cells of the culture were largely tetraploid (Wright & Northcote, 1973).

The cell has evolved an alternative technique for making more 'master copies' in order to meet the heavy demands for RNA production involved in cytodifferentiation. This is the functional significance of endoreduplication. Nagl (1973) demonstrated by tritiated uridine autoradiography in *Allium* that RNA synthesis continued through endomitosis, in contrast to mitosis, and that the amount of RNA synthesized was doubled from one level of ploidy to the next higher one. This uninterrupted RNA synthesis during endoreduplication may be a necessity for the completion of certain steps in cytodifferentiation, and such RNA synthesis might be suppressed by the events of mitosis (Nagl, 1973). In an examination of the changes in nuclear DNA content in *Cymbidium* protocorms cultured *in vitro* and treated with various growth regulators, Nagl & Rücker (1972) found that 2,4-D treatment produced some giant nuclei in differentiating xylem elements which exhibited polytenic structures at approximately the 128C level. Differentiating trichoblasts exhibited endoreduplication and contained considerably more cytoplasmic and nuclear RNA than neighboring nondifferentiating cells (Cutter & Feldman, 1970). D'Amato (1965) considered that endopolyploidy exerts an influence on cytodifferentiation by 'intensifying the action of specific loci in the chromosomes'. An interesting question, however, is why we find such a high variability in the ploidy levels of differentiating tracheary elements (Phillips & Torrey, 1974). A target cell may require a sufficient amount of a specific functional RNA in order to produce a protein necessary for the initiation of the CDS, and this specific RNA may only be formed at some period of the cell cycle. The cell must either remain within the critical period for a sufficient length of time in order to produce the required level of RNA, or it must recycle and return to this point again for further synthesis of the particular RNA fraction. The ploidy level of the differentiating cell will be inversely related to the efficiency of the cell in synthesizing the 'cytodifferentiation RNA'. A diploid cell, held in the critical period long enough for a sufficient amount of RNA production, will shortly exhibit signs of cytodifferentiation.

We must not lose sight of the fact that in some systems there is a strict control of the sequence of DNA synthesis leading to mitosis, with cytodifferentiation only at the diploid (2C) level. Brunori & D'Amato (1967) examined the embryos of dry seeds of *Pinus pinea* and *Lactuca sativa* and found only 2C nuclear DNA contents in the embryonic root cells of both plants. Since differentiated provascular cells were already present in the mature embryos, we must conclude that these cells had differentiated at the 2C level.

Gene amplification

The differential replication or amplification of part of the genome results in the production of enormous numbers of specific genes in certain developmental processes (see Ebert & Sussex, 1970). Gene amplification has been suggested to play a role during cytodifferentiation in plants (Avanzi & D'Amato, 1970; Avanzi, Cionini & D'Amato, 1970; Innocenti & Avanzi, 1971; Nuti Ronchi, 1971; Nagl, Hendon & Rücker, 1972; Avanzi, Maggini & Innocenti, 1973). If gene amplification occurs during cytodifferentiation, the DNA content of the nuclei should be magnified, but not in multiples of the 2C level. Innocenti & Avanzi (1971) examined the labeling of differentiating metaxylem in *Allium* with tritiated thymidine, tritiated uridine and tritiated lysine. Chromosome endoreduplication was followed by gene amplification in the nucleolar-organizing region which contained cistrons for coding rRNA. Labeling experiments demonstrated that these nucleoli synthesized DNA, RNA and proteins. Avanzi *et al.* (1973) employed a mixture of 18 S and 25 S tritiated rRNA fractions for hybridization with DNA in differentiating metaxylem elements of *Allium*, and labeling was detected over nucleolus-associated DNA. With the DNA–rRNA technique, these workers indicated that DNA extracted from root tissues, when extra DNA synthesis occurred in differentiating metaxylem elements, contained approximately sixfold rDNA compared with the DNA extracted from undifferentiated meristematic cells. (The term 'rDNA' refers to 'ribosomal DNA' or genes coding for rRNA.) Experiments with *Cymbidium* have shown that exogenous growth regulators are capable of enhancing or inhibiting DNA amplification (Nagl & Rücker, 1972; Nagl *et al.*, 1972). It is possible that different tissues in various organisms have solved the problem of RNA production during cytodifferentiation by alternative methods. In the cortical cells of the pea root, in which endoreduplication and cytodifferentiation were induced, the DNA content of the nuclei followed multiples of 2C with no indication of gene amplification (Phillips & Torrey, 1974).

Other nuclear DNA

The presence of a metabolically labile DNA, subject to rapid turnover, has been reported in the cytodifferentiation zone of the roots of *Vicia* (Sampson & Davies, 1966). The DNA content of the differentiating cells of *Vicia* possibly involves metabolic DNA (Brunori, 1971). A similar view has been expressed from microphotometric studies of *Pisum* roots by Van Parijs & Vandendriessche (1966). The concept of 'metabolic DNA' has been discussed by Iwamura (1966) and Granick & Gibor (1967). The suggestion that the synthesis of a special type of DNA may be necessary for cytodifferentiation has been mentioned by several animal physiologists. Lasher (1971) has presented evidence of a special type of DNA synthesis necessary for the enhancement and stabilization of chondrogenesis in several cartilage-producing tissues, and it may or may not be associated with S-phase DNA synthesis. The synthesis of this special type of DNA was considerably more sensitive to various levels of thymidine monophosphate than the S-phase DNA. Similar findings have also been reported by Cahn (1968) and Holtzer & Abbott (1968).

The following summary indicates our present tentative position:

1. Determination probably occurs either during or following DNA replication, and this marks the initiation of the CDS. DNA synthesis is a necessity for cytodifferentiation.
2. Cytodifferentiation *may* occur, under special circumstances, following DNA synthesis in the absence of cell division. Parenchymatous cells of *Camellia* leaf destined to become photosynthetic mesophyll cells exhibited nuclear activity and were evidently reprogrammed without cell division. This probably occurs to some extent in the cytodifferentiation of *Coleus* wound vessel members and in the formation of tracheary elements under in-vitro conditions.
3. Cells in G_1 show visible signs of tracheary element differentiation, but evidence is not conclusive that the appearance of cytodifferentiation is strictly confined to G_1; G_2 remains a possibility. The functional significance of G_0 in this process remains to be elucidated.

4. The progressive development of the different stages of the CDS occurs at varying rates, under different in-vitro conditions, and results in a variety of sizes and shapes of differentiated cells. The possible relationship of cell expansion to the cell cycle has not been studied.

5. The following hypothesis may be offered concerning the significance of the DNA synthesis requirements for cytodifferentiation. The demands for the production of specialized proteins for the various stages of the CDS require the operation of efficient transcriptional and translational processes. Different tissues and different organisms may have evolved different methods in order to achieve these synthetic demands by either (*a*) normal cell cycle operation, (*b*) endoreduplication or polyteny, (*c*) gene amplification, or (*d*) some other special type of DNA production.

There is a serious need for more information on the cytodifferentiation relationships of auxin, cytokinin, gibberellin and ethylene to the different periods of the cell cycle. Various combinations of hormones may alter the timing of the cell cycle by either lengthening or shortening different periods. Future studies should consider several questions. Which 'cytodifferentiation proteins' signal the beginning of the CDS, and is their synthesis confined to a particular period of the cell cycle? Do plant hormones, acting as critical variables, operate at the transcriptional or translational level in the regulation of the synthesis of these proteins? Are these newly synthesized enzymes involved in either the formation or deposition of secondary wall materials?

5

Regulation of secondary xylem differentiation

The tissues produced by the vascular cambium are among the most complex found in plants (Philipson, Ward & Butterfield, 1971), and the cytodifferentiation processes occurring within the cambial zone have been exceedingly difficult to study and interpret. Although this discussion will focus on the hormonal regulation of secondary xylem differentiation, some closely related aspects of cambial physiology should be examined. What are the hormonal factors involved in the resumption of cambial activity from the dormant state? How does the process sustain itself throughout the growing season? What is known concerning the cessation of division and cytodifferentiation at the end of the growing season? Endogenous hormones released under natural growing conditions stimulate cell division and the subsequent cytodifferentiation of the newly formed derivatives, yet these two processes can be experimentally separated by the production of 'undifferentiated' initials (Shininger, 1970, 1971). In other words, it would appear that the stimuli necessary for cell division and cytodifferentiation are not exactly identical.

Primary xylem elements differentiate from procambial cells determined in the apical meristems, whereas secondary xylem elements originate from recently divided cells of the vascular cambium. The terminology of Wilson, Wodzicki & Zahner, (1966) seems appropriate for our discussion of the cambial region: (*a*) the cambial zone composed of the immediate derivatives of the cambial initial and mother cells capable of further mitoses, (*b*) the contiguous zone of radially enlarging derivatives no longer dividing, (*c*) the zone of maturing derivatives where secondary wall thickening occurs, and (*d*) the mature derivatives. When cambial initials divide, one of the daughter cells becomes either a xylem or a phloem cell and the other cell remains meristematic. Two kinds of xylem derivatives are produced: the fusiform initials that differentiate to form tracheary elements and the ray initials that form ray parenchyma cells (see Esau, 1965*a*). Fusiform initials are elongated with tapering ends, whereas the ray initials are isodiametric and relatively small. Sanio (1873) demonstrated that cells develop in groups of four each within the cambial zone and this early observation has been confirmed (see Timell, 1973). Each group of 'Sanio's four' (Mahmood, 1968) develops by two successive divisions of a single cambial initial. Contiguous with and on the inside of the group of Sanio's four we find a similar cluster of enlarging tracheary elements termed 'the enlarging four' (Mahmood, 1968).

The differentiating tracheary elements of the secondary xylem pass through the same ontogenetic changes that are associated with vessels and tracheids of the primary xylem; i.e., cell enlargement, secondary wall thickening and pit formation, lignification, and finally autolysis of the cell contents with maturation. Secondary wall thickening in primary xylem elements usually occurs in an annular, spiral, or scalariform-reticulate pattern, whereas in secondary xylem, definite layers (S_1, S_2, S_3) are deposited on the inner surface of the primary wall.

The physiology of secondary xylem differentiation has been extensively reviewed, and only selected aspects of cytodifferentiation will be examined in this chapter (Brown, 1970; Denne, 1970; Philipson *et al.*, 1971; Zimmermann & Brown, 1971; Kozlowski, 1971; Thimann, 1972; Steeves & Sussex, 1972).

A working hypothesis concerning seasonal reactivation of the vascular cambium in forest trees gradually evolved among developmental botanists. Expanding buds in the early spring release substances that initiate mitotic activity in the cambial zone. Cell division activity gradually proceeds basipetally from these expanding buds to the basal regions of the plant. Cambial activity spreads rapidly in species with ring-porous wood, rather slowly in species with non-porous wood, and most slowly in species with diffuse-porous wood. In ring-porous hardwoods, whose name is derived from the ring-like appearance of the large vessels in transverse section, the vessels produced in the early wood are considerably larger in diameter than those produced later in the growing season (Fig. 20). In conifers the wood is relatively homogeneous, i.e., non-porous, and it is

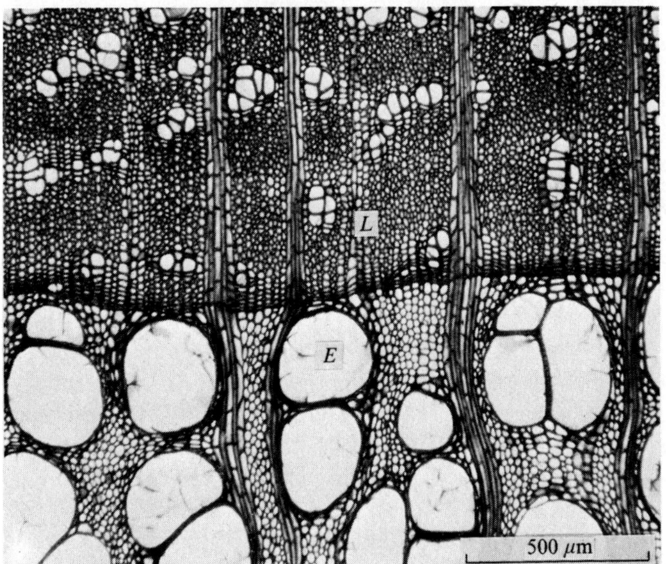

Fig. 20. Transverse section of a ring-porous hardwood (*Morus rubra* L., red mulberry) showing a greater distribution of vessels in the earlywood (*E*) than in the latewood (*L*). Relatively large vessels are formed only during the development of the earlywood.

composed mainly of tracheids, ray parenchyma cells and resin canals. When the vessels are approximately equal in diameter, and more or less uniformly distributed throughout the annual growth ring, the wood is termed diffuse-porous (Fig. 21). It was clear from early studies that the rate of cytodifferentiation in the newly activated cambial zones was different for trees of these three different types (Lodewick, 1928; Priestley, Scott & Malins, 1933, 1935). In ring-porous trees the first newly formed xylem was coincident with, or just prior to, the breaking of buds, whereas xylem differentiation in diffuse-porous species occurred some two to three weeks later, at which time the leaves were from one-quarter to fully expanded (Lodewick, 1928).

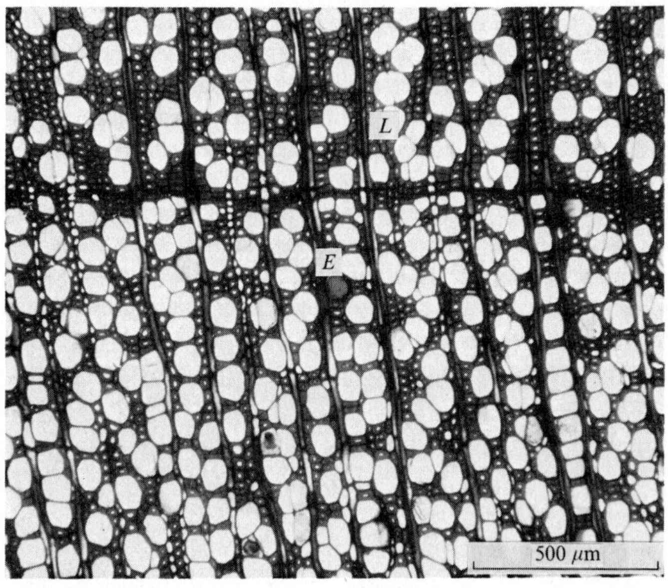

Fig. 21. Transverse section of a diffuse-porous hardwood (*Liquidambar styraciflua* L., sweetgum) showing a relatively uniform distribution of vessels in both latewood (*L*) and earlywood (*E*).

The rapidity of the seasonal reactivation in the ring-porous species is intriguing. This is thought to be due to the presence of an auxin precursor in the bark of these trees (Wareing, 1951; Digby & Wareing, 1966). This precursor, probably tryptophan, is presumably converted into auxin at the time of bud break simultaneously at all levels down the tree. Szalai & Gracza (1958) reported that free tryptophan was present in the bark and buds of the ring-porous species *Fraxinus excelsior*. In *Fraxinus* the stimulus for the initiation of cambial activity moved basipetally and was blocked by girdling, but renewed cambial activity was observed some distance below the girdle (Wareing, 1951). The stimulus for the latter activity was associated with the positions of adventitious buds, although these buds had been removed at the beginning of the experiment. In contrast to these results with *Fraxinus*, cambial activity was invariably associated with active bud growth in the diffuse-porous species *Acer pseudoplatanus* and *Tilia europaea* (Wareing, 1951). Similarly Reines (1959) found that cambial activity was blocked by the disbudding of *Prunus serotina* (diffuse-porous), but was unaffected by the same treatment in *Carya glabra* (ring-porous).

Some of the techniques that have been employed in experimental studies on cambial activity have included: (*a*) surgery on intact trees, e.g., disbudding, girdling, and isolation of bark patches, (*b*) feeding of isolated internodal stem segments, and (*c*) extraction and identification of hormones and other substances from cambial tissues. Several difficulties are encountered with the use of the disbudding method. Disbudded woody stems cannot be presumed devoid of all regulatory substances, and traces of auxins and other hormones are usually present. The application of a single growth regulator may be sufficient to stimulate cambial activity in the presence of minute quantities of endogenous hormones (Wareing *et al.*, 1964). Disbudding may be partially ineffective because of the presence of adventitious buds hidden within the bark. These can be sources of auxin since they may be released from dormancy by the disbudding surgery (Wareing, 1951; Zimmermann & Brown, 1971). Nevertheless, several groups of workers have employed disbudding techniques, and we will examine these results in a subsequent section involving hormonal interactions (see pp. 61–2).

The results obtained by Brown (1970) on the breaking of dormancy in isolated internodal stem segments may have involved the presence of minute amounts of endogenous growth substances within the tissues at the time of removal. These experiments were based on feeding the isolated segments a sterile culture medium containing growth regulators (Brown & Wodzicki, 1969). Some species formed cambial derivatives on the xylem side without exogenous auxin, whereas other species showed no mitotic activity following applications of IAA or kinetin, either separately or in combination (Brown, 1970). Zajaczkowski (1973) reported that isolated stem segments of *Pinus silvestris* continued to differentiate secondary xylem for several weeks of culture provided that sucrose and IAA were present in the medium, but that the isolated segments showed an inexplicable seasonal variation in their responsiveness to the auxin treatment.

The surgical isolation of circular patches of bark was employed by Evert & Kozlowski (1967) as a technique for the study of cambial reactivation and cytodifferentiation in a diffuse-porous species, trembling aspen (*Populus tremuloides* Michx.). Two concentric rings were cut into the bark, and all of the tissues external to the secondary xylem were removed between the concentric rings. The wood surface, from which the bark had been removed, was scraped, and the gap was filled with grafting wax. The treatment was devised in order to prevent the conduction of phloem assimilates to the isolated tissues. When the bark was isolated during dormancy, a resumption of cambial activity and phloem differentiation occurred in the spring. However, xylem differentiation was completely inhibited by bark isolation during the dormant season. Bark isolation after xylem differentiation had commenced in the spring resulted in the formation of relatively short tracheary elements with poorly thickened secondary walls. The ultimate effect of bark isolation, either during dormancy or growth, was the differentiation of all fusiform initials to parenchymatous cells (Evert & Kozlowski, 1967). Evidently the hormonal requirements for the initiation of cambial activity and phloem differentiation, in contrast to xylem differentiation, are somewhat different. In a subsequent study, Evert and his co-workers (1972) examined the effects of the surgical isolation of bark patches on another diffuse-porous species, sugar maple (*Acer saccharum* Marsh.). Bark isolation during the dormant season of this species prevented cambial activity and cytodifferentiation in isolated areas in approximately one-half of the trees, although varying amounts of abnormal cytodifferentiation were observed in the remainder of the treated trees. Isolation of the bark patches after the resumption of cambial activity and phloem differentiation, but before the initiation of xylem formation, produced narrow xylem increments composed of atypically-short vessel members and parenchyma cells. Bark isolation after xylem formation had commenced seriously affected secondary wall deposition in the newly formed tracheary elements (Evert *et al.*, 1972). Both of these studies by Evert and his co-workers clearly demonstrated that normal differentiation of secondary xylem elements is dependent on a continual supply of phloem assimilates, at least in these diffuse-porous species.

Another experimental approach has involved extraction and identification of the endogenous hormones present in the various tissues of the cambial zone. Digby & Wareing (1966) studied the distribution of substances with auxin activity in a ring-porous (*Ulmus glabra*) and a diffuse-porous (*Populus trichocarpa*) species during the spring. An auxin precursor was present at each of the three stem heights sampled in the ring-porous species prior to bud swelling. The precursor remained present in significant amounts at the time of bud swelling and three weeks after bud break. On the other hand, in the diffuse-porous tree neither auxin precursors nor auxin activity was found at any of the three stem heights sampled prior to bud swelling. Only the uppermost level showed auxin activity coinciding with the time of bud swelling, and three weeks later auxin activity was present at all three stem levels (Digby & Wareing, 1966). Sheldrake (1971) extracted substances from the cambium and from differentiating xylem and phloem in three hardwoods (*Acer pseudoplatanus* L., *Fraxinus excelsior* L., and *Populus tremula* L.) during the growing season, and he reported that most of the substances exhibiting auxin activity were identified on chromatograms as IAA. The concentration of these auxin-like substances was highest in the differentiating xylem and least in the differentiating phloem. Whitmore (1968) examined cambial extracts of *Populus deltoides* and *Pinus sylvestris* and reported that they contained substances with the same chromatogram R_f value as IAA and were active in the wheat coleoptile bioassay. The major growth substance detected by mesocotyl bioassay in extracts of cambial tissues of *Pinus radiata* was identified as IAA by chromatography, electrophoresis, Salkowski test, and absorption and fluorescence spectroscopy (Sheperd & Rowan, 1967). A relatively high concentration of indolic growth-promoting substances in the cambial zone of *Larix decidua* was present during earlywood differentiation, and the cytodifferentiation of latewood correlated with reduced amounts of the indolic compounds with increasing concentrations of phenolic growth inhibitors (Balatinecz & Kennedy, 1968). Zajaczkowski (1973) claimed that the spring initiation of cambial activity could not be correlated with any consistent concentration gradient of natural auxin extracted from the cambial region of *Pinus silvestris* trees, and he stressed the importance of the responsiveness of the cambial tissues to the hormones. Wodzicki & Wodzicki (1973) were unable to detect any seasonal decrease of concentration of extractable auxin from the cambial region of *Pinus silvestris* trees, and they reported that the activity of extracted pine auxin had different characteristics from IAA as determined by standard auxin bioassays. Because of the inconsistency of these results with cambial extracts, it becomes apparent that physiological activity in the cambial zone is regulated by a complex system of factors that extends beyond the presumed presence or absence of indolic auxins.

During the 1930s the demonstration of the physiological properties of IAA led to the hypothesis that the basipetal transport of endogenous IAA produced during bud activation was the hormonal stimulus initiating cambial divisions in the spring (Zimmerman, 1936; Söding, 1937; Avery *et al.*, 1937). Tepper & Hollis (1967) examined the movement of the stimulus for mitotic activity in reactivating terminal buds of white ash (*Fraxinus americana* L.). Cell divisions were initiated in leaf primordia and mitotic activity progressed to the procambium and then to the cambium at the base of the terminal bud. Divisions of the cambium were then observed in progressively more proximal regions of the shoot system. Whether or not this wave of mitotic activity reflected auxin movement is a debatable point. The spring initiation of cambial activity in adult trees of *Pinus silvestris*, under natural conditions, was *not* correlated with any concentration gradient of natural auxin extracted from the cambial zone of these trees (Zajaczkowski, 1973). The relationship between auxin concentration and cambial activity is complex, and it involves alterations in the responsiveness of the cambial cells. Zajaczkowski (1973) was unable to relate the interaction of other plant growth regulators and sucrose, respectively, with the seasonal variation in the cambial response to auxin.

The problem of cytodifferentiation of the derivatives is immensely complicated, since the process involves hormone and metabolic gradients in space and time, i.e., both horizontally and longitudinally in differentiating secondary tissues of varying physiological and chronological ages. The inaccessibility of the developing tissues poses special problems in technique. From the standpoint of cytodifferentiation, the most important critical variables are auxin, cytokinins, gibberellins, sugar nucleotides and possibly ethylene. It seems probable that these critical variables change during the CDS. Controlling variables may include temperature, water stress, internal pressures, light and gaseous components of the atmosphere. The immediate source of the hormonal and metabolic requirements for cytodifferentiation is not clearly understood. Hormones and other chemical variables could be arriving at the sites of xylem differentiation from (*a*) the translocation stream of the mature phloem, (*b*) the transpiration stream of the mature xylem conducting root-synthesized hormones, (*c*) the final stages of cytodifferentiation of the tracheary elements and sieve elements, and (*d*) the meristematic activity of the dividing cells of the cambial zone (Fig. 22). Once the cambium has commenced mitotic activity, it may generate its own source of auxin and possibly other growth regulators, independent of a continuous supply from either developing or mature shoots (Sheldrake, 1973*a*). This would mean that cytodifferentiation, once it is initiated, then becomes self-sustaining or autocatalytic. Jacquiot (1950) demonstrated that cambial explants of several different woody species were capable of sustained cambial activity on culture media for six to eight weeks without a supply of exogenous auxin.

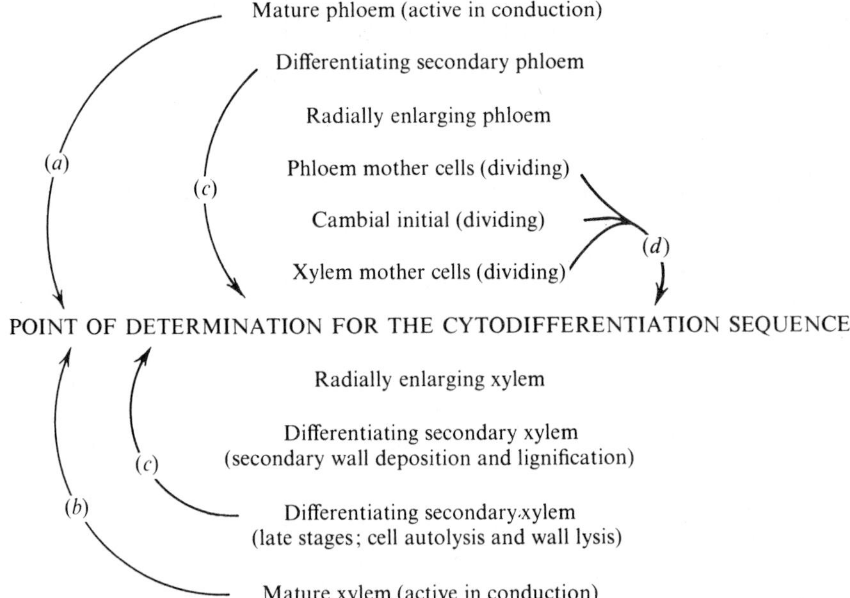

Fig. 22. Diagram of the arrangement of differentiating secondary vascular tissues indicating the possible sources of xylogenic hormones. The point of determination for the CDS of a tracheary element probably occurs either during or shortly after the division of the xylem mother cell. Endogenous plant hormones, operating as critical variables in the CDS, may move laterally into the CDS zone from (*a*) the translocation stream of the mature phloem, (*b*) the transpiration stream of the mature xylem, (*c*) the terminal stages of cytodifferentiation of secondary xylem and/or phloem elements, and (*d*) the meristematic activity of the cells of the cambial zone.

Sheldrake (1973*a*) has suggested that auxin is produced as a consequence of cell death. According to Sheldrake (1973*a*), tryptophan is maintained at too low a concentration in living cells for the production of auxin, but autolysis of differentiating xylem cells results in a release of tryptophan which then may be converted to auxin by adjacent living cells. Sheldrake (1973*a*) has reviewed the evidence that gibberellins, cytokinins and ethylene probably arise from dying cells during tracheary element differentiation.

We cannot discount the possibility that physiologically active auxin may be transported in the phloem and move laterally into the zone of xylem differentiation. Several investigators have indicated that auxin moves in the phloem (Peel, 1964, 1966; Eschrich, 1968; Lepp & Peel, 1971*a,b*). The movement of indole auxins in *Salix viminalis* L. occurred in the sieve tubes (Hoad, Hillman & Wareing, 1971). Bonnemain (1971) found that auxin from mature leaves of *Vicia faba* moved via the phloem, whereas auxin from immature leaves was conducted by the cambium and differentiating xylem.

Several investigators have given evidence to support the hypothesis that the normal functioning of the cambial zone involves the interaction of several different types of hormones. Auxins and gibberellins together produced a synergistic effect when applied in lanolin paste to disbudded shoots of *Acer, Populus, Robinia* and *Fraxinus* (Wareing, 1958; Wareing *et al.*, 1964; Digby & Wareing, 1966). The application of auxin alone induced the formation of xylem elements in discontinuous groups around the stem periphery, and gibberellic acid alone caused cell division throughout the vascular cambium with the production of an inner layer of secondary xylem derivatives which failed to differentiate into normal xylem elements. The combination of auxin and gibberellin stimulated the formation of a relatively wide zone of secondary xylem composed of normally developed vessels, fibers and tracheids. A comparable role of exogenous gibberellic acid in stimulating cambial activity in gymnosperms has not yet been demonstrated, and these plants may have endogenous gibberellins in supra-optimal amounts for cambial processes (Brown, 1970). The delay in bud activity of Douglas-fir (*Pseudotsuga menziesii*) seedlings because of low soil temperatures was, according to Lavender *et al.* (1973), eliminated by exogenous gibberellin treatment. These workers found a positive correlation between increased bud activity, level of gibberellin-like compounds in the xylem sap, and elevated soil temperatures. The application of IAA or NAA in lanolin paste to the cut surfaces of disbudded *Pinus silvestris* induced the formation of a ring of secondary xylem the entire length of five-year-old trees (Hejnowicz & Tomaszewski, 1969). The application of *cis-* and *trans-*cinnamic acids, coumarin, L-tryptophan, kinetin, benzyladenine, and gibberellic acid, respectively, did not induce cambial division. However, exogenous gibberellic acid (GA) and cytokinins, given simultaneously with either IAA or NAA, accelerated the basipetal stimulus which was induced by the auxins, and resulted in normal xylem development. Similar results were obtained by Robards, Davidson & Kidwai (1969) with disbudded stem segments of *Salix*. They stressed the importance of a simultaneous treatment of IAA, kinetin, GA and sucrose. This induced more than twice as much cytodifferentiation and nearly fifty per cent more cell production, when compared with control segments. The relative effects of young and mature leaves on cambial activity and secondary xylem differentiation was examined in bean (*Phaseolus vulgaris* L.) seedlings by Hess & Sachs (1972). Quantitative data showed that young leaves could be imitated, in terms of xylem formation, by a source of exogenous auxin, whereas xylem differentiation induced by mature leaves was only mimicked by a combination of auxin and gibberellin (Hess & Sachs, 1972).

A disturbed hormonal balance may be responsible for the anatomical characteristics of a disease of peach called 'stem pitting' (Whitmore & Jones, 1972). In diseased trees the cambial initials in the root collar region were relatively short and differentiated into parenchymatous cells instead of normal tracheary elements. Hormonal analyses indicated that the cambial tissues of the diseased trees were abnormally high in gibberellins and low in auxin. A condition similar to 'stem pitting' was induced by treating dormant twig sections with GA (100 mg/l), whereas normal secondary xylem formation appeared with the addition of IAA (10 mg/l) to the GA (Whitmore & Jones, 1972).

The synergism that exists between auxin and gibberellin in cytodifferentiation may involve auxin transport. In willow stem segments the presence of GA in the secondary xylem increased the lateral mobility of labeled IAA and IAA-aspartic acid, and the effect was enhanced by the exogenous application of GA to the xylem (Field, 1973). This finding may be related to the remarkable effect of GA on the pattern of IAA-induced xylogenesis in explants of Jerusalem artichoke tuber (Dalessandro, 1973*b*).

The possible interrelationships that may exist between phloem and xylem differentiation have not been examined critically. Early work on the initiation of cambial activity led to the hypothesis that xylem differentiation in the spring generally preceded that of phloem (Evert, 1963). Recent studies have indicated that in many species phloem differentiation precedes that of xylem by as much as several weeks. In *Pyrus communis* and *Pyrus malus* phloem differentiation commenced six to eight weeks before xylem formation (Evert, 1960, 1963); in *Pinus strobus,* four weeks (Alfieri & Evert, 1965); and in *Populus tremuloides,* six weeks (Davis & Evert, 1965). Since the initiation of phloem differentiation does not appear to be co-ordinated with bud opening, some interesting questions arise. What is the stimulus for phloem differentiation? Does the differentiation of phloem derivatives influence the subsequent CDS of the xylem elements?

Virtually nothing is known concerning the physiology of the cessation of cambial activity. Reinders-Gouwentak (1941) suggested that inhibitory substances may play roles in the regulation of cambial activity. Some workers have hypothesized that a continuous supply of diffusible auxin from actively growing shoots and leaves is a necessity for continued cambial activity, and that the cambial zone itself is incapable of supplying its own source of auxin and other hormonal needs (Zimmermann & Brown, 1971). Although this concept has merit, the production of abscisic acid by the mature leaves may play an important role in the inhibition of cambial division and cytodifferentiation. Hess & Sachs (1972) found that abscisic acid greatly reduced the differentiation of new xylem whether it was applied alone or in combination with other growth regulators. Balatinecz & Kennedy (1968) suggested that the accumulation of phenolic compounds in the xylem was associated with the cessation of cell division and cytodifferentiation.

The period of time required for cambial initials to complete the CDS leading to the production of mature secondary xylem tissue has been examined in *Pinus radiata* (Skene, 1969), *Pinus banksiana* (Kennedy & Farrar, 1965) and *Pinus resinosa* (Whitmore & Zahner, 1966). In *Pinus radiata* the average time for the cambial initials to complete a division cycle was approximately four weeks, with little evidence of variation during the growing season. Radial enlargement, prior to the appearance of the S_1 layer of secondary wall, required three weeks early in the season. The time of radial enlargement gradually decreased to ten days later in the season. Approximately three to four weeks were required for the deposition of the secondary wall, and this period increased to about eight to ten weeks later in the season (Skene, 1969). It was estimated that the entire sequence of tracheary element differentiation was completed in three weeks in seedlings of *Pinus banksiana* (Kennedy & Farrar, 1965). In *Pinus resinosa* it was estimated that radial enlargement required three to five weeks followed by three to six weeks for maturation (Whitmore & Zahner, 1966). It seems remarkable that the phases of secondary xylem differentiation *in vivo* are extended to such long periods of time when compared with the relative rapidity of the same events in cultured explants. If the events in cytodifferentiation *in vitro* could be experimentally arrested at various stages, perhaps by withholding certain metabolites at critical times, this technique might prove valuable for future studies.

The experiments of Shininger (1970) have given us an interesting insight into the timing of cytodifferentiation, since he demonstrated that secondary xylem differentiation in *Xanthium* can be examined as an event independent of cambial division. Cambial division occurred in the absence of leaves and in the absence of exogenous growth regulators, whereas normal secondary xylem differentiation required either the presence of expanding leaves or exogenous growth regulators. The removal of leaves and buds from the shoots of *Xanthium* seedlings caused the immediate cessation of xylem fiber and tracheid differentiation with the continued production of cambial derivatives; potential fibers and tracheids became thin-walled parenchymatous cells. Normal small vessels continued to be produced from the cambial derivatives. Apparently vessel differentiation was directly related to the production of vessel initials by the cambium; the percentage of cells formed by the cambium which developed into vessels was the same in intact, decapitated, or single-leaf plants. The development of a single leaf from a lateral bud or a decapitated plant provided sufficient stimulus to restore temporarily fiber and tracheid formation only during the time of rapid expansion of the leaf blade. It was suggested that immature leaves provided growth factors for the elongation, secondary wall formation and lignification of potential tracheids and fibers (Shininger, 1970). When a single leaf was permitted to develop in a decapitated plant, a transient 'flush' of fiber–tracheid growth factors was produced. The most recently formed cambial derivatives differentiated into fiber–tracheids; derivatives present at the time of decapitation remained parenchymatous. Shininger (1970) considered two possible explanations: (*a*) the first-formed derivatives were 'cut off' from the supply of growth factors by the late-formed derivatives that differentiated, or (*b*) the first-formed derivatives were cut off *in time* from the possibility of xylogenesis by some inherent capability. The most interesting question raised by Shininger's study concerns the uniqueness of fiber–tracheid differentiation. Why should vessel differentiation be less demanding in terms of growth factors then fiber–tracheid formation? A clue may be found in the recent work of Morey (1973) on the effects of DPX-1840, an auxin transport inhibitor, on altering xylem differentiation in *Prosopis* (See p. 108). In the increment of secondary xylem formed during DPX-1840 treatment, only a few cambial derivatives differentiated as fibers, and most derivatives formed thick-walled axial parenchyma cells. However, the inhibitor treatment was without effect on the percentage of vessel members that differentiated during the same period (Morey, 1973).

The hormonal, nutritional and biophysical factors governing the inception of a vascular cambium *de novo* in cultured plant tissues are poorly understood (Gautheret, 1966; Torrey, 1966), and the requirements cannot be compared with the inception of a vascular cambium in an interfascicular tissue zone (Siebers, 1971*a*,*b*). The types of cytodifferentiation occurring as a result of cambial activity *de novo* will depend, to some extent, on the origin of the explant system (Jacquiot, 1957; Gautheret, 1966) and the composition of the culture medium (Wetmore & Rier, 1963). Gautheret (1966) pointed out that the type of explant tissue had some unknown 'orienting' influence. Explants of phloem initiated cambia at the periphery and produced secondary phloem toward the interior and exterior xylem, but the relative positions of the secondary tissues were reversed with xylem explants. The cambium that arises *de novo* in a callus may take different forms: it may form an ill-defined meristematic zone; it may develop a structure somewhat akin to a vascular bundle; or it may be confined to a relatively few divisions in a vascularized nodule. In any case, it has been impossible to duplicate the complex organization of secondary tissues characteristic of the intact plant by experimental manipulations with tissue explants (Torrey *et al.*, 1971). It is impossible to excise a fragment of a cambial region and induce it to continue its unique activity *in vitro* without external pressure. Such isolated cambia may be induced to continuous division, but the result invariably is a parenchymatous callus (Steeves & Sussex, 1972). Nevertheless, the resulting callus may subsequently form cambia and secondary tissues (Saussay, 1969).

One neglected area of study is the role of pressure and cell confinement on the cytodifferentiation of secondary xylem elements (Brown & Sax, 1962; Brown, 1964). In young trees of *Populus trichocarpa* and *Pinus strobus*, longitudinal bark strips were lifted away from the bole and enclosed with plastic film to prevent desiccation (Brown & Sax, 1962). A callus pad formed on the inner surface of the exposed strip, and subsequently a new phellogen and cambium differentiated and extended around the outer periphery of the callus pad. With continued cell division, the progressively higher mutual cell pressures led to a more orderly and normal cytodifferentiation of xylem elements (see Chapter 8). In another experiment the bark strips, encased in polyethylene plastic film, were pressed back into the wound site and held in position with externally applied pressure. After a slight amount of cell proliferation by ray cells, bringing the entire inner surface under more-or-less uniform pressure, the parenchyma cells differentiated as xylem elements. The cambium functioned normally, and cytodifferentiation of the derivatives was completely normal (Fig. 23). A more striking

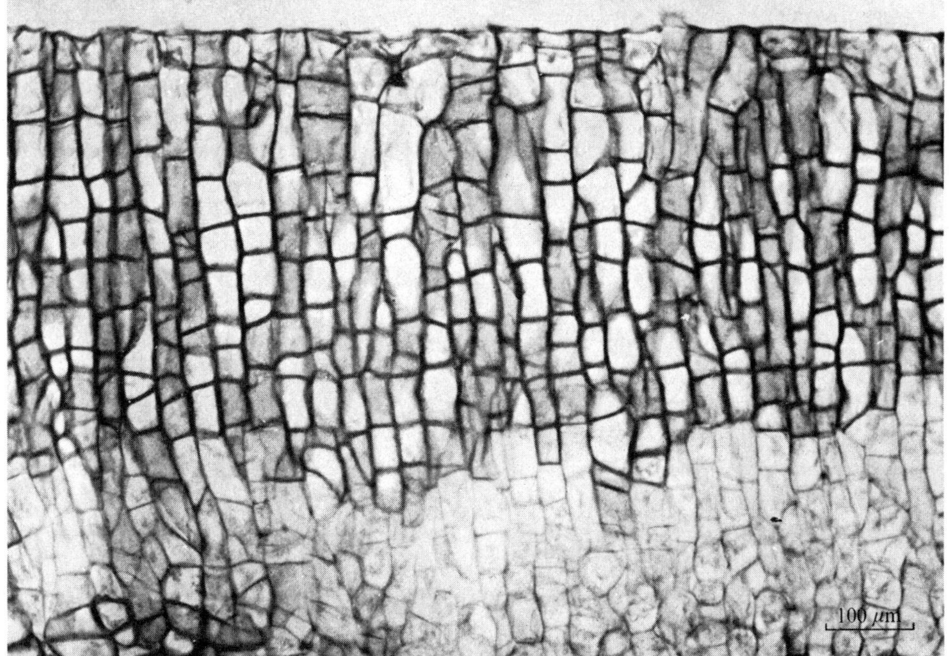

Fig. 23. Transverse section of a callus pad formed on the inner surface of a bark strip of *Populus trichocarpa*. The bark strip had been lifted from the bole, encased in plastic, and permitted to grow for 14 days without external pressure. The bark strip, with the newly formed callus pad, was then pressed back into the wound cavity with an external pressure of 0.5 atm. The callus pad was subjected to external pressure for an additional 30-day period. The pressure induced some form of cytodifferentiation in the outermost parenchyma cells of the callus. (From Brown & Sax, *Am. J. Bot.* **49,** 1962.)

demonstration of the importance of pressure in cytodifferentiation was given by Brown (1964). Explants, consisting mainly of inner phloem and the cambial region, were prepared from eastern cottonwood (*Populus deltoides*) and cultured on a nutrient medium containing IAA and kinetin. The resulting cell proliferation produced an extensive unorganized callus mass with a complete loss of cambial identity. However, with the application of external pressure, cytodifferentiation of xylem elements was induced. Relatively low pressures (0.05 atm) were sufficient to induce distinct changes in cytodifferentiation. It seems clear from these experiments that pressure *alone*, without any change in the hormonal or nutritional status of the culture medium, can bring about secondary wall thickening in callus parenchyma (Brown, 1964). A possible explanation may involve the pressure-induced release of ethylene by the stressed cells (Goeschl, Rappaport & Partt, 1966). Experiments similar to those conducted by Brown (1964) should be repeated with an attempt to correlate pressure application, ethylene release and cytodifferentiation.

Excised roots, grown aseptically in culture solutions, lack a vascular cambium (Torrey, 1963). However, a vascular cambium in isolated radish roots was initiated by a basal feeding of auxin, cytokinin and sucrose (Loomis & Torrey, 1964; Torrey & Loomis, 1967*a*,*b*). The response was further enhanced by the addition of *myo*-inositol (Fig. 24). The cytodifferentiation of secondary xylem according to specific cell types was not subject to experimental control (Torrey & Loomis, 1967*a*,*b*). A similar study of excised radish roots, cultured with a basal application of auxin and cytokinin, was made by Webster & Radin (1972). The pericycle divided initially, and subsequent division was observed in the procambium with large numbers of secondary derivatives produced at both sites of division. Cultured roots treated with auxin alone produced only a limited number of divisions of the pericycle. In roots treated with cytokinin, but without auxin, all of the cells of the pericycle divided, and a multiseriate zone of pericycle-derived cells developed. The procambium was not markedly affected by the application of either of the hormones singly. Cambial activity in radish roots is probably regulated by zeatin and its derivatives (Radin & Loomis, 1971). Extraction of cytokinins from radish roots showed the presence of zeatin, zeatin ribonucleotide, zeatin ribonucleoside and an unidentified cytokinin unrelated to either zeatin or isopentenyladenine (Radin & Loomis, 1971). Previously, Torrey (1963) demonstrated that isolated pea roots required only auxin and sucrose for cambial formation and suggested that these roots probably contain sufficient amounts of endogenous cytokinins for the initiation of a cambium. The regulation of cambial activity in seedling roots and in isolated cultured root tips of turnip (*Brassica rapa*) was examined by Peterson (1973). The excision of the root tip had no apparent effect on cambial activity in the seedlings, whereas excision of the shoot system greatly reduced both the number of cambial derivatives and the differentiation of secondary xylem elements. Removal of either one or two cotyledons, on otherwise intact plants, reduced cambial activity in the root. Cul-

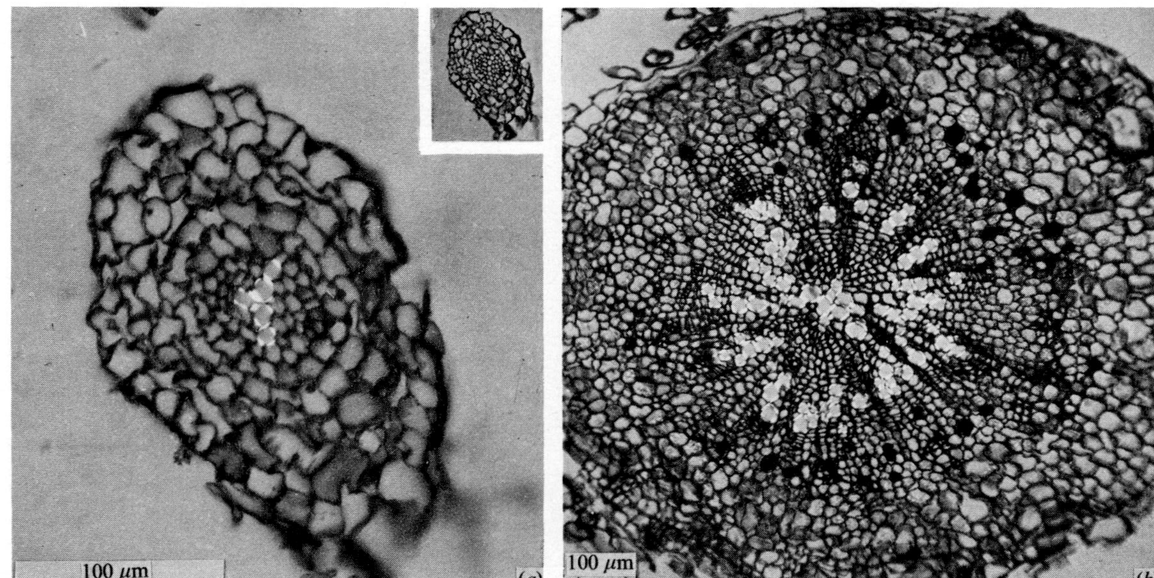

Fig. 24. Transverse sections about 10 mm from the base of second-passage radish (*Raphanus sativus* L.) roots after 20–30 days of culture. Polarized light was employed to show the birefringent secondary walls of the tracheary elements. (*a*) A root grown on a basal nutrient medium without exogenous growth regulators. (Inset is same magnification as *b*.) (*b*) With a basal application of IAA (10^{-5} M), *myo*-inositol (100 mg/l) and 6-benzylaminopurine (1 mg/l) cambial activity was observed. (From Loomis & Torrey, *Proc. Nat. Acad. Sci.* (USA), 1964.)

tured first-transfer root tips required the presence of exogenous amounts of a combination of auxin, cytokinin, sucrose and *myo*-inositol for maximal radial growth (Peterson, 1973).

Flower buds of *Asclepias* were cultured in order to determine the relative effects of auxin and cytokinin on the secondary growth in the pedicels (Safwat, 1969). The initiation and activation of the cambium was achieved on media containing a combination of IAA and kinetin, and it was suggested that these hormones stimulated cambial activity during fruit development in this plant.

Caruso & Cutter (1970) reported that the shoot system of a leafless mutant of tomato produced no vascular cambium, yet a vascular cambium was induced in the hypocotyl of the mutant by grafting an actively growing normal shoot tip to it. The authors suggested that the results be interpreted in terms of the normal shoot tip acting both as a physiological sink, with respect to cell-division stimuli produced mainly in the root system, and as a physiological source, with respect to cambial-activating hormones.

The inception of cambial activity in interfascicular areas of the stem of castor bean (*Ricinus communis* L.) raises an interesting question (Siebers, 1971*a*,*b*, 1972). Is the initiation process dependent on inductive 'influences' arising from the adjacent vascular bundles, or is the process autonomic in the sense that the initiation has been predetermined at an early age of shoot development? Siebers (1971*a*) surgically removed blocks of interfascicular tissue, rotated them 180°, and implanted them back in the incision cavities. Cambial activity occurred at the original site within the blocks; xylem and phloem were formed at the topographically outer and inner side of the tissue blocks, respectively. There was no evidence to suggest the existence of an inductive effect imposed by the established cambium of the vascular bundles. In another experiment, interfascicular tissue blocks were removed for culturing from hypocotyls of six-, seven-, or eight-day-old castor bean seedlings (Siebers, 1971*b*). In the intact seedlings, interfascicular cell division and cambium formation was observed to commence on the eleventh or twelfth day. Since cambial activity proceeded normally in approximately one-half of the interfascicular explants cultured on a basal medium containing mineral salts and sucrose, Siebers (1971*b*) concluded that the established vascular tissues of the seedlings do not supply the parenchymatous cells of the interfascicular region with a specific inductive factor for mitosis and cytodifferentiation. Exogenous IAA and kinetin induced the formation of a high percentage of disoriented cambial walls. Indoleacetic acid induced xylogenesis in some parenchyma cells of the pith, which resulted in the differentiation of tracheary elements of noncambial origin. A high concentration of IAA (12.5 mg/l) induced xylogenesis directly in the cell layer destined to form the cambial layer without prior tangential cell division (Fig. 3). When combinations of plant hormones were employed, the effect of GA–kinetin produced the greatest number of tracheary elements and the least disorientation of cambial walls. Only rarely were tracheary elements of noncambial origin formed during the latter treatment (Siebers, 1971*b*). Interfascicular blocks cultured on a liquid medium containing mineral salts, sucrose and kinetin also produced a vascular cambium (Siebers, 1972).

The location of initiating factors governing the most distal vascular cytodifferentiation in the shoot apices of vascular plants have generated considerable debate among plant morphologists; this discussion has been outlined in a recent publication by McArthur & Steeves (1972). The shoot apex of *Geum chiloense* possesses a short cylinder of tissue which presumably represents provascular tissue (McArthur & Steeves, 1972). Surgical elimination of leaf primordia by a puncture technique (Wardlaw, 1947; Soe, 1959) suggested that the initial stage of vascular cytodifferentiation, which precedes procambial development, was completely under the influence of the terminal meristem. The application of exogenous IAA and sucrose to isolated shoot apices without leaf primordia enhanced the development of the procambial cylinder below the provascular tissue (McArthur & Steeves, 1972).

In resumé, the hormonal stimuli involved in the seasonal resumption of cambial activity are poorly understood, and the phenomenon involves endogenous factors that are, as yet, unknown. Because of the unusual location of the zone of cytodifferentiation, there are four possible sources of critical and controlling factors: assimilates from the mature phloem, differentiating phloem elements and cambial derivatives, autolyzing tracheary elements, and substances from the transpiration stream. In addition to 'tissue competence' there are reasons to believe that cytodifferentiation involves auxin, gibberellins, cytokinins, carbohydrates and possibly ethylene. The importance of pressure in cytodifferentiation suggests that stress-induced ethylene may be a factor. Experiments involving the surgical isolation of bark revealed that cambial activity and phloem differentiation are, in some manner, less demanding in terms of requirements than xylem differentiation. Xylogenesis apparently requires the presence of some leaf- or shoot-derived substances available from the phloem translocation stream, and these factors probably include auxin and carbohydrates. There are indications that the requirements for the cytodifferentiation of vessels are not the same as the requirements for fiber–tracheid formation. Maximal development of a vascular cambium in isolated first-transfer roots of both radish and turnip was achieved with a culture medium that included auxin, cytokinin, sucrose and *myo*-inositol, and this may be a general requirement for cambial development in root tips. Hormonal requirements for division and cytodifferentiation of a presumably 'predetermined' interfascicular cambium include GA and cytokinin, but probably not auxin.

6

Ultrastructural studies of differentiating xylem elements

Ultrastructural investigations conducted during the 1960s on the cytodifferentiation of tracheary elements have been reviewed previously (Roberts, 1969; Robards, 1970; O'Brien, 1974). The present discussion, however, focuses on dilemmas arising from ultrastructural observations. Differences exist between the mature cells comprising the primary and secondary xylem. Consequently, cytological observations may differ somewhat concerning these two tissues. The primary xylem elements possess secondary walls deposited in distinctive patterns of unique ridges, whereas the secondary xylem consists of a more or less uniform secondary wall deposited on the inner surface of the primary wall with the exception of pit fields. In addition, the secondary wall of the secondary xylem is formed in successive layers, i.e., S_1, S_2, and S_3. It is important to bear in mind that various fixation procedures have been employed in the studies under discussion, and that the results obtained following the destructive effects of potassium permanganate cannot be compared reliably with cells fixed in combinations of glutaraldehyde and osmium tetroxide.

Primary xylem

Although there has not been a critical examination of any cytochemical and ultrastructural differences that may exist between procambial and nonprocambial cells in the apical meristem (Esau, Cheadle & Gill, 1966*a*,*b*; O'Brien, 1974), procambial cells are morphologically distinguishable in the dome of the quiescent center of roots (L. J. Feldman, personal communication). Observations were recently made on the ontogeny of vessel elements from procambial cells in leaves of *Zea mays* L. (Srivastava & Singh, 1972) and in root tips of *Medicago sativa* (Maitra & De, 1971). Initially, the enlargement of the procambial cells was accompanied by increased vacuolar volume, but shortly afterwards increased cytoplasm synthesis occurred and only a few small vacuoles were observed. The cells destined for cytodifferentiation as xylem elements were characterized by an abundance of rough-type endoplasmic reticulum, polyribosomes, proliferating mitochondria, and dictyosomes with numerous vesicles, in contrast to adjacent stelar cells at the same locus (Maitra & De, 1971; Srivastava & Singh, 1972). The procambial cells of *Zea* contained amorphous masses of fibrillar material which disappeared during secondary wall formation (Srivastava

& Singh, 1972). The observations by Sinnott & Bloch (1944, 1945) with the light microscope on the 'prepatterning' of the localization of the future secondary wall thickenings by underlying bands of densely granular cytoplasm have not been supported by most ultrastructural investigations (Wooding & Northcote, 1964; Cronshaw, 1965*a*,*b*; Esau *et al.*, 1966*a*,*b*; Pickett-Heaps & Northcote, 1966*a*; Hepler, Fosket & Newcomb, 1970; Srivastava & Singh, 1972). Goosen-De Roo (1973*a*), however, examined the fine structure of the surface of the protoplast of differentiating tracheary elements of cucumber after plasmolysis and observed cytoplasmic ridges in the areas originally located between the wall thickenings.

Numerous workers have observed microtubules localized directly over the developing ridges of secondary wall, and the concept has developed that microtubules are somehow involved in the orientation of the newly formed cellulose microfibrils (Maitra & De, 1971; Srivastava & Singh, 1972; see reviews by Newcomb, 1969; Pickett-Heaps, 1974; Hepler & Palevitz, 1974). Some have suggested that microtubules may channel vesicles filled with wall materials into the developing thickenings (Pickett-Heaps, 1968; Robards, 1968; Newcomb, 1969; Northcote, 1969*b*; Maitra & De, 1971), provide structural support for particles at the plasmalemma–cell wall interface (Murmanis, 1971*b*), or conduct fluids (McManus & Roth, 1965). A possible function has been suggested by the abnormal distribution of microtubules in the xylem of apple trees infected with 'rubbery wood' disease (Nelmes, Preston & Ashworth, 1973). In this pathological condition, fewer microtubules were present, lignification was incomplete in the xylem fibers and vessels, and the xylem cells were abnormally thickened and coarsely textured. The entire shoot system of the infected plants was highly flexible. The percentage of cellulose in 'rubbery wood,' as determined by chemical methods, was the same as in normal apple wood (Scurfield & Bland, 1963). These observations suggest that microtubules may be involved in the deposition of lignin precursors in the cell wall, and that the bonding between lignin and cellulose microfibrils gives rigidity to the secondary xylem (Nelmes *et al.*, 1973). Colchicine has been used to study microtubular function, since this alkaloid binds to a subunit of microtubular protein and thus prevents the assembly of microtubules (Borisy & Taylor, 1967*a*,*b*). Tracheary elements differentiated in the presence of colchicine have a unique secondary wall pattern; instead of sharply defined angular bands of secondary wall, the ridges have a wavy, undulating appearance (Pickett-Heaps, 1967; Roberts & Baba, 1968*a*; Hepler & Fosket, 1971; Figs 6, 25). This wavy pattern was evidently a reflection of the swirled appearance of the cellulosic microfibrils as seen with the electron microscope (Hepler & Fosket, 1971). Thus, microtubules appear to affect neither the synthesis nor the deposition of cellulosic mircrofibrils in the secondary wall, but only the orientation of the microfibrils. This view was confirmed by the observation that both the degree of polymerization and yield of cellulose in *Valonia* cells were essentially the same either in the presence or absence of colchicine (Marx-Figini, 1971). Hepler

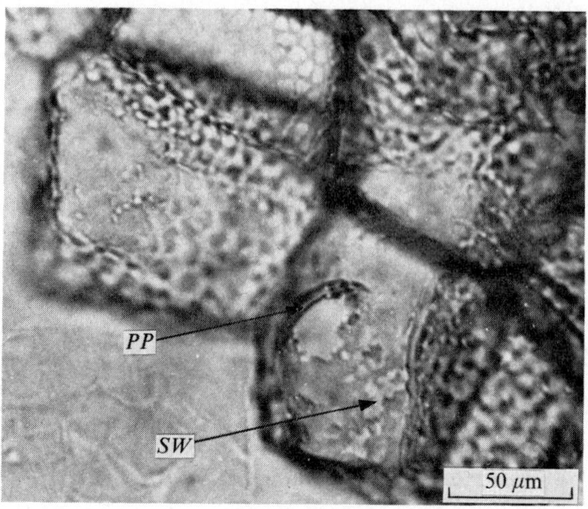

Fig. 25. Effect of colchicine on secondary wall deposition in differentiating tracheary elements. Explants of lettuce pith parenchyma, cultured for 7 days on a xylogenic medium containing colchicine (0.04 %), contained tracheary elements with irregular secondary walls (*SW*). Note what appears to be a partially completed simple perforation plate (*PP*).

& Fosket (1971) observed that even in the absence of microtubules in colchicine-treated tissue the secondary thickenings of two adjacent xylem elements were formed directly opposite one another across a common primary wall (Fig. 26). The formation of perforation plates was evidently unaffected by the colchicine treatment (Hepler & Fosket, 1971), but the pitting pattern of xylem elements of *Zea* was altered by a similar treatment (Stein, Rowley & Lockhart, 1971). In colchicine-treated pea roots, metaxylem elements failed to differentiate normally, but the protoxylem was unaffected (Barlow, 1969*a*). The metaxylem elements were abnormally large, unlignified and thin-walled. Similar xylem abnormalities were found in pea roots after the incorporation of high levels of tritiated thymidine, i.e., metaxylem was abnormal and protoxylem was unaffected (Hummon, 1962); these abnormalities were not observed when an undetermined labeled fraction from tritiated thymidine was incorporated into the developing secondary wall of tracheary elements in young *Xanthium* leaves (Maksymowych, 1973). Earlier, Eigsti (1938) reported that xylem differentiation was capable of continuing in roots of *Zea* and *Allium* grown for extended periods of time in the presence of concentrations of colchicine as high as 0.1 per cent. Since colchicine has certain side effects, e.g., on ethylene biosynthesis (Noodén, 1971) and on membrane structure (Wunderlich, Müller & Speth, 1973), this alkaloid may affect cytodifferentiation in a complex manner. How microtubules might regulate microfibril orientation is perplexing, because the microtubules, located in the cytoplasm, are separated from the developing wall by the plasmalemma (Hepler & Fosket, 1971; Robards & Kidwai, 1972). Some function may be served by cross-

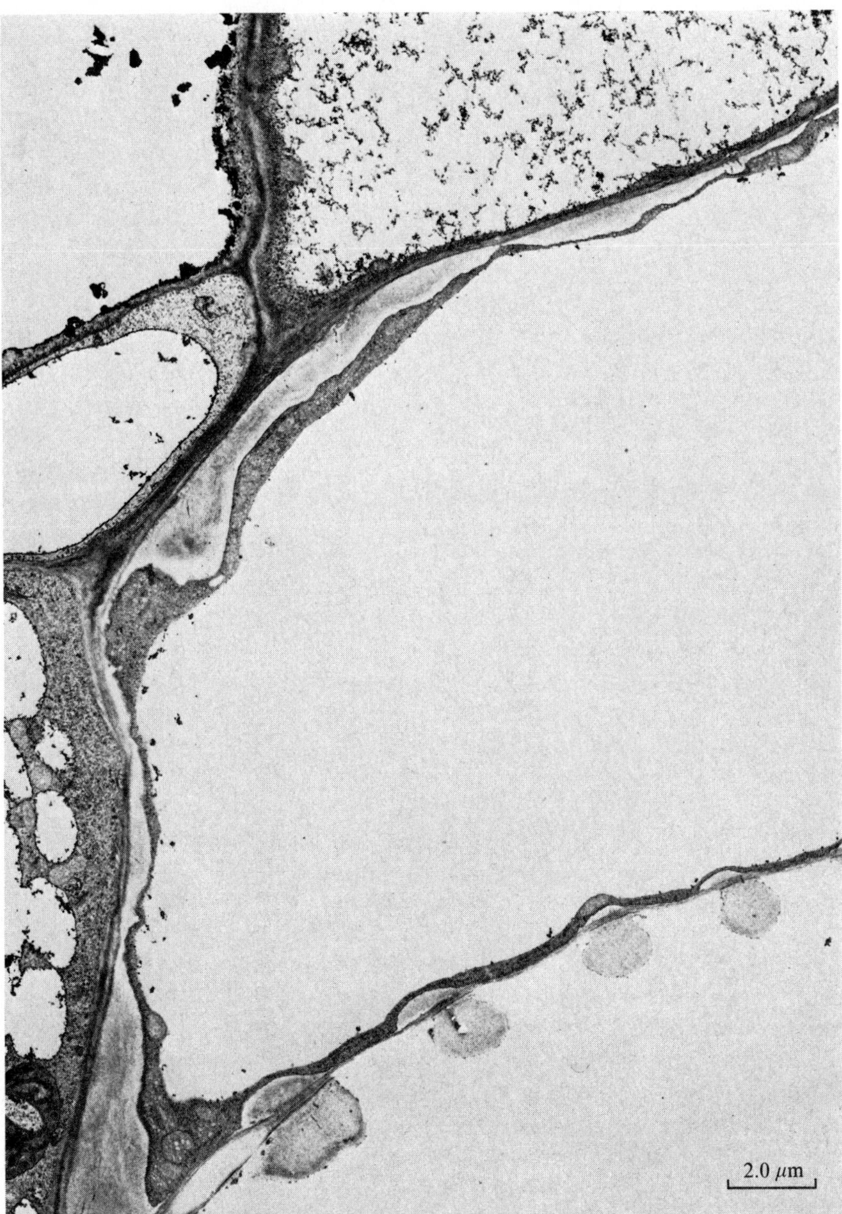

Fig. 26. Ultrastructural view showing secondary wall formation in a vessel member of *Coleus* during culture of the stem segment in the presence of colchicine. Along the lower side of the central cell the wall thickenings occurred in relatively well-defined ridges opposite the wall striations of the mature tracheary element. On all other sides, however, the differentiating central cell was bordered by parenchyma cells, and secondary wall formation was spread over the entire surface of the primary wall. (From Hepler & Fosket, *Protoplasma* **72,** 1971.)

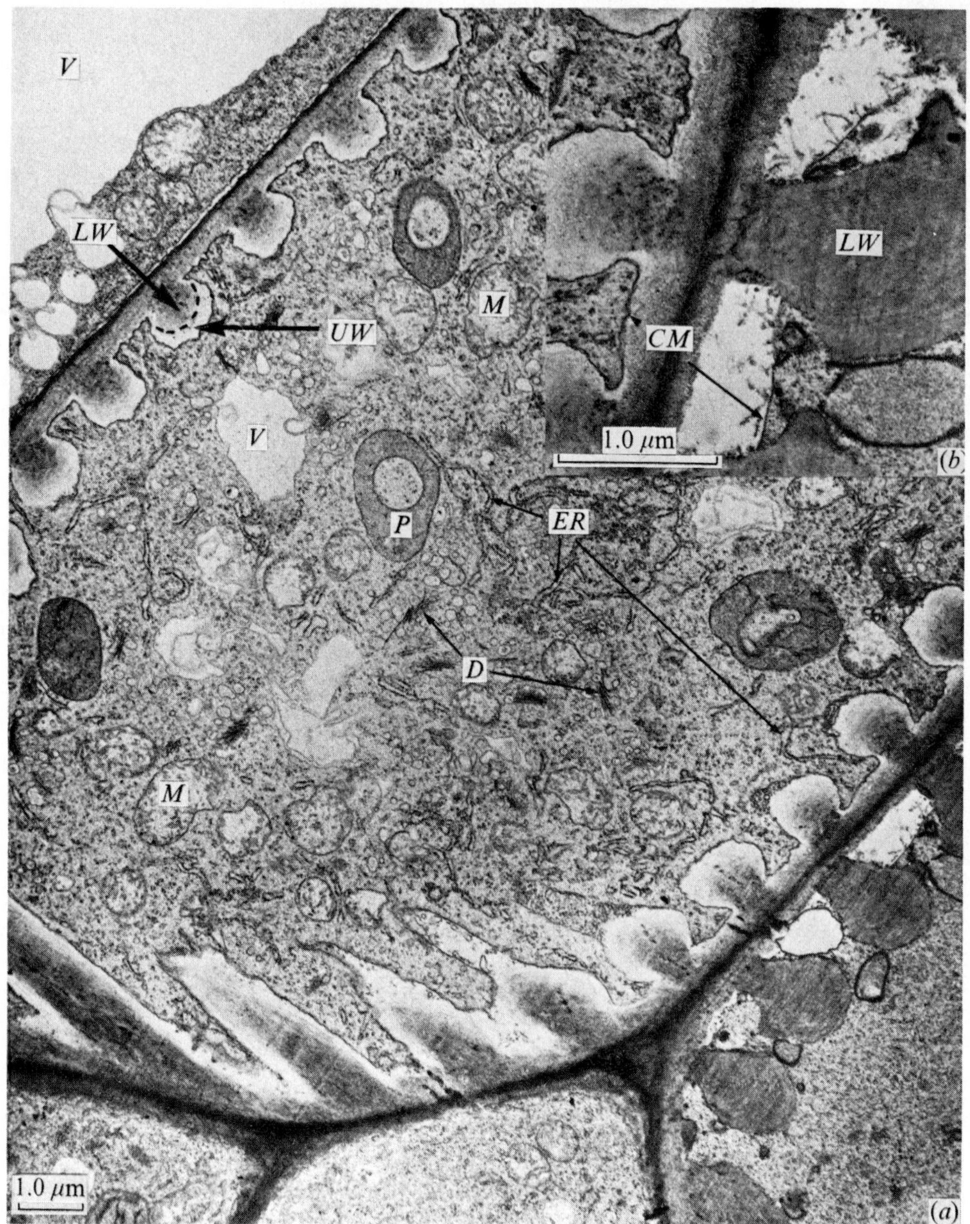

Fig. 27. Ultrastructural views of stages in the cytodifferentiation of tracheary elements in a leaf of *Phaseolus vulgaris* L.

(*a*) The differentiating cell is forming secondary wall thickenings and lignification has commenced. Dictyosomes (*D*), rough endoplasmic reticulum (*ER*) and mitochondria (*M*) are abundant during this developmental stage. *LW*, lignified wall; *UW*, unlignified wall; *P*, plastid; *V*, vacuole.

bridges between closely adjacent tubules, between microtubule and vesicle, or between microtubules and some membranous component of the cell (see Hepler & Palevitz, 1974). Cronshaw (1967) observed cross-bridges between microtubules and the plasmalemma at sites where the membrane covered the developing secondary wall in differentiating tracheary elements. The physiology of microtubules has been reviewed (Tilney, 1971; Pickett-Heaps, 1974; Hepler & Palevitz, 1974).

A functional connection may exist between organelles during wall formation. Maitra & De (1971) state: 'the frequent occurrence of vesicles completely encircled by microtubules indicates that it is not a case of chance association of the two organelles, but that it is a distinct design by which the microtubules trap the free-moving vesicles and finally the vesicles are deposited along the cell wall.' O'Brien (1974) has pointed out some of the problems in attempting to relate cytoplasmic fine structure to a particular event in cytodifferentiation. It is comparatively easy to demonstrate that differentiating tracheary elements contain an abundance of various organelles and a highly ordered distribution of microtubules, but the structure of a static cell surface section cannot be established without an analysis of serial sections or an examination of stereoscopic pairs, and these approaches have not been employed in studies on differentiating xylem elements (O'Brien, 1974). The abundance of cytoplasmic organelles during tracheary element differentiation is evident (Fig. 27), but the dynamic functional relationships between the organelles and the cell wall cannot be fathomed from a few random micrographs.

What, if anything, is transported to the developing secondary wall by dictyosome vesicles or other cytoplasmic organelles is still a matter for speculation. Certain wall-localized enzymes, e.g., β-glucan synthetase (Ray, Shininger & Ray, 1969) and peroxidase (Hepler, Rice & Terranova, 1972) were reported to be localized in dictyosome vesicles; perhaps these enzymes were directed to certain sites in the developing wall by microtubules. However, experimental evidence suggesting the association of either the dictyosome vesicles or the endoplasmic reticulum with secondary wall components has met with criticism (Torrey *et al.*, 1971; O'Brien, 1972, 1974). Chafe & Wardrop (1970) employed a freeze-etching technique and found no correlation between the distribution of particles on the surface of the plasmalemma and the microfibril orientation in differentiating xylem fibers of *Eucalyptus*. The incorporation of labeled cinnamic acid, phenylalanine and glucose in differentiating tracheary elements has been followed by microautoradiography (Wooding, 1968; Pickett-Heaps, 1968). Both cinnamic acid and phenylalanine are precursors of lignin; glucose serves as a

---

(*b*) A later stage of cytodifferentiation showing a withdrawal of the cell membrane (*CM*) from the wall. Part of this cell can be seen at lower magnification in the lower right-hand corner of (*a*). *LW*, lignified wall. (From O'Brien & McCully, *Plant structure and development. A pictorial and physiological Approach,* 1969, Macmillan. Courtesy of the authors and publisher.)

monomer for polysaccharides. According to O'Brien (1972), 'though label may be incorporated into dictyosomes or ER in cells depositing secondary wall, there is no evidence that the label in these organelles is in wall material or that these systems contribute membrane-bound label to the wall.' A contrary view has been expressed with the suggestion that pectins and hemicelluloses are packaged in Golgi-derived vesicles and exported to the wall through the plasmalemma by reverse pinocytosis (Robards, 1970; see review by Northcote, 1971). Recently Fowke & Pickett-Heaps (1972) examined cell wall deposition in fiber cells of *Marchantia* and reported that vesicles and cell wall were intensely stained following peroxidation with silver hexamine, indicating the presence of polysaccharides. The vesicles were presumed to be Golgi-derived. When immature thalli were fed labeled glucose, much of the label was associated with Golgi bodies. If a cold chase followed the incubation period with the label, the label was localized primarily over the cell wall (Fowke & Pickett-Heaps, 1972).

Although there is debate concerning the possible roles of the endoplasmic reticulum, electron micrographic evidence is lacking in regard to any generalization about its cytological location (O'Brien, 1972). Several workers have indicated that the ER appeared to be localized along regions of the wall not producing secondary thickenings, and it has been suggested that the ER may either prevent secondary wall formation or channel wall substances to adjacent areas of deposition (Pickett-Heaps, 1966, 1967; Pickett-Heaps & Northcote, 1966*a*; Northcote, 1969*b*). Burgess & Northcote (1968) suggested that the smooth-type ER may be associated with the transport and aggregation of subunits of the microtubules.

During the maturation of a tracheary element there is a progressive autolysis of the protoplast and hydrolysis of selected portions of the primary wall. Factors involved in the initiation of autolysis are unknown, and the onset of maturation may be delayed for a lengthy period of time during cytodifferentiation in secondary xylem (Skene, 1969; see Chapter 5). We may speculate on some possible reasons for the relative rapidity of the maturation of primary xylem elements. The differentiating cell may experience autotoxicity because of an over-production or under-utilization of lignin metabolites. These products of secondary metabolism doubtless are utilized and redistributed more efficiently within the orderly files of secondary xylem elements. Another possibility is that at some point during cytodifferentiation there is either the release, synthesis, or activation of cell wall hydrolyases and lysosomes (Gahan & Maple, 1966; see review by Gahan, 1973), and these enzymes are toxic to the cell. This is supported by the observation that general autolysis and wall hydrolysis appear to occur simultaneously (O'Brien, 1970). From a study of the hydrolyzed walls in the xylem of *Phaseolus vulgaris* L. and *Triticum aestivum* L. leaves, O'Brien (1970) suggested that such localized digestion is probably a general phenomenon in the maturation of all tracheary elements. Xylem and phloem parenchyma cells contiguous with differentiating tracheary elements may provide some protection to the common wall layers from digestion by hydrolyases released during autolysis (O'Brien, 1970). Hydrolyases degrade all areas of the cell wall unprotected either by lignin or some unidentified middle lamella component (O'Brien & Thimann, 1967; O'Brien, 1970; Srivastava & Singh, 1972). This hydrolysis involves the removal of polyuronides, pectins and hemicelluloses, leaving a considerably weakened cellulosic residue (O'Brien & Thimann, 1967). According to Czaninski (1972*b*), the presence of lignified thickenings is not an indispensable factor in preventing the hydrolysis of the primary wall in vessels of *Robinia* and *Acer*. The origin of these hydrolyases is of great interest. Czaninski (1972*b*) has suggested that they may arise, at the time of cytoplasmic lysis, either from vessel-associated cells or from transported xylem sap. The complete removal of wall regions, e.g., in the formation of the simple perforation plate, may be partly mechanical under the influence of transpirational flow (O'Brien, 1974). Bierhorst & Zamora (1965) indicated that lignified areas of the wall can be hydrolyzed in the formation of perforation plates in some plants. It is difficult, however, to understand how lignin is capable of conferring protection from enzymatic hydrolysis. Since cellulase has been found in the guttation fluid of several coleoptiles (Sheldrake & Northcote, 1968*a*; Sheldrake, 1970), we would expect the hydrolyzed walls and pit membranes consisting of cellulose to be completely removed by the cellulase activity (O'Brien, 1974). The dissolution of cell wall material has been observed in provascular elements of the quiescent root meristems of *Allium* bulbs; the breakdown process was associated with the presence of endoplasmic reticular elements that may transport hydrolyases to the wall (Bal & Payne, 1972). A similar view was expressed by Srivastava & Singh (1972). Gamaley (1971) observed structures resembling animal cytosegresomes and autolytic vacuoles in differentiating protoxylem elements of *Pinus*.

Secondary xylem

Relatively few studies have been made on the ultrastructural cytology of secondary xylem elements during cytodifferentiation, and critical distinctions have not been made between actively dividing cambial initials and radially enlarging cambial derivatives on the xylem side. The first indication of cytodifferentiation in active cambial cells is a marked enlargement of the derivatives, achieved mainly by a radial expansion of the cells (Wilson, 1963; Cronshaw, 1965*a*; Srivastava & O'Brien, 1966; Czaninski, 1968, 1970). Many small vacuoles observed in the cambial initial disappear with the concomitant formation of a large central vacuole. At this stage of development the cytoplasm and organelles form a thin layer along the cell wall. Shortly thereafter, the deposition of cell wall material is accompanied by increased activity of the endoplasmic reticulum, dictyosomes and plasmalemma (Kollmann & Schumacher, 1964; Cronshaw, 1965*a*; Srivastava & O'Brien, 1966; Czaninski, 1968, 1970; see review by Catesson, 1974). Plasmalemma invaginations, packed with microfibrillar material of the same nature as

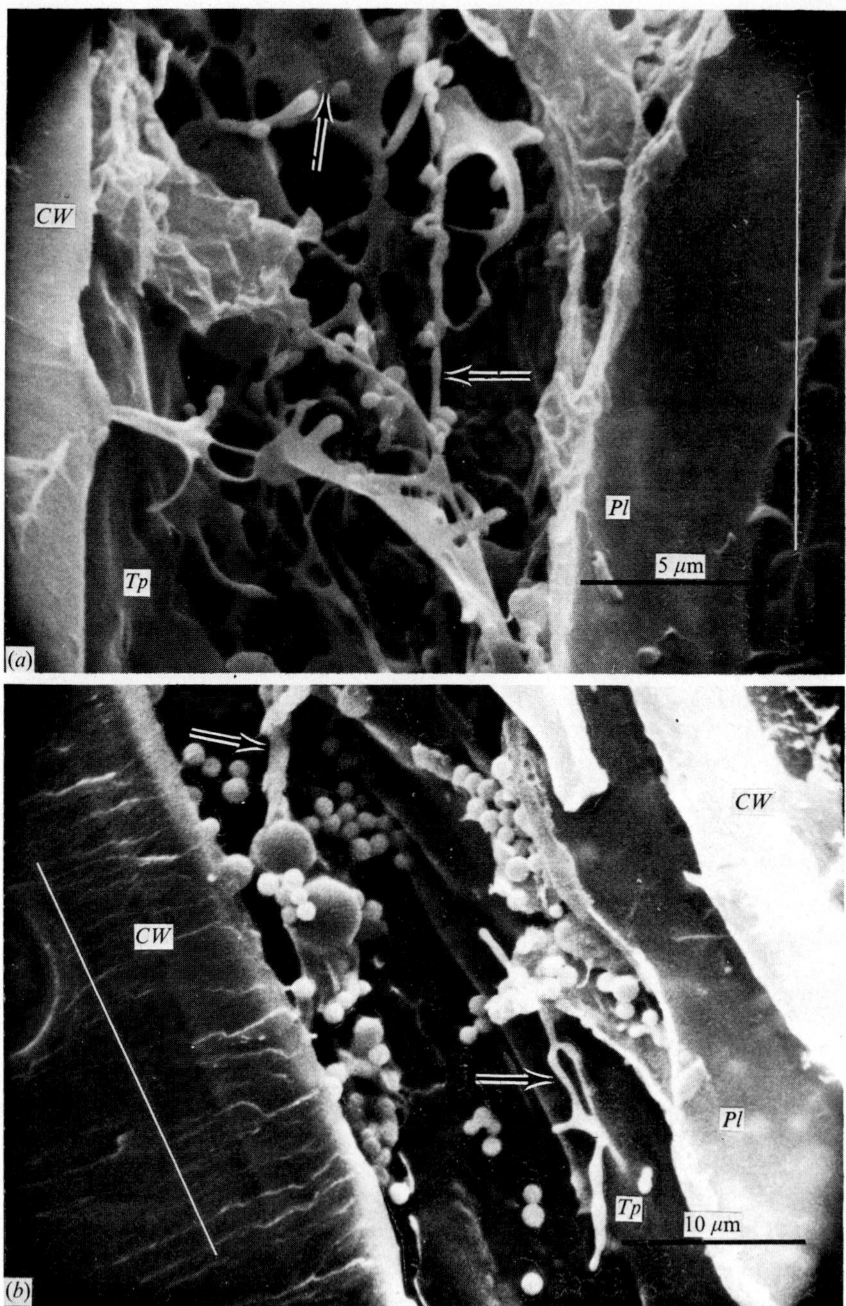

Fig. 28. Organization of intravacuolar cytoplasm in maturing tracheid of *Pinus echinata* Mill. as revealed by scanning electron microscopy. Radial longitudinal view after a portion of the cell wall and outer layer of cytoplasm were fractured away. Longitudinal axis of the differentiating cell is indicated by the white line. Intravacuolar cytoplasm is indicated by arrows. *CW*, cell wall; *Pl*, plasmalemma surface; *Tp*, tonoplast surface.

that constituting the cell wall, can be seen (Robards, 1968; Timell, 1973). Lomasomes, as described by Mahlberg, Olson & Walkinshaw (1970, 1971), have been observed in differentiating xylem cells of *Pinus strobus* (Murmanis & Sachs, 1973) and *Picea abies* (Timell, 1973). These lomasomes contain membrane-bound vesicles that appear to be filled with a fibrillar material (Timell, 1973). Ultrastructural differences between dormant and active cambial cells have been described by numerous workers (Srivastava & O'Brien, 1966; Robards & Kidwai, 1969*b*; Mia, 1970; Itoh, 1971; Barnett, 1973; Timell, 1973). The ultrastructure of stem cambia in arborescent plants has been extensively reviewed by Catesson (1974).

Robards & Kidwai (1972) examined cytodifferentiation in xylem fibers of *Salix*. This cell type was chosen mainly because of a gelatinous wall layer which served as a useful visual marker. These fibers were characterized by an abundance of dictyosomes and associated vesicles which these workers assumed were engaged in hemicellulose production. Microtubules were consistently parallel to cellulose microfibril orientation in all stages of development, yet the relative density of microtubules varied considerably from cell to cell. At three or four positions along each fiber, usually near the center and at the tips, the major organelles were closely grouped together in the ground-plasm. These organelle clumps were separated by large vacuoles bordered with a peripheral layer of cytoplasm. The differentiating fibers were rich in rough endoplasmic reticulum and the nuclei had deeply infolded envelopes. Paramural bodies (Marchant & Robards, 1968), observed earlier in differentiating xylem of willow (Robards, 1968), were completely absent from these gelatinous fibers (Robards & Kidwai, 1972). Although this description of the ultrastructure of xylem fibers of *Salix* is interesting, our knowledge of cytodifferentiation is not greatly enriched by it.

Murmanis and co-workers have examined cytodifferentiation in the secondary xylem of *Pinus strobus* L. (Murmanis & Sachs, 1969; Murmanis, 1970, 1971*a*,*b*). During the early stages of cytodifferentiation, the enlarging fusiform derivatives were indistinguishable from cambial initials with regard to organelle composition (Murmanis & Sachs, 1969). The ray system consisted of ray parenchyma cells and tracheids. The ray tracheids, at an early stage of differentiation, were distinguishable from ray parenchyma cells by the presence of bordered pits. The secondary walls and pit borders of ray tracheids were thinner than those of contiguous axial tracheids and they lacked the S_2 layer of the secondary walls of axial tracheids (Murmanis & Sachs, 1969). Murmanis (1971*b*) reported that particles in the plasmalemma–cell wall area, presumably engaged in wall synthesis, were closely associated with microtubules in differentiating cells of *Pinus*.

(*a*) Tracheid during cytodifferentiation prior to autolysis of protoplasm. Intravacuolar cytoplasm consists of complex filamentous reticulum immersed in vacuolar sap.

(*b*) Final stage of maturation indicated by breakdown of intravacuolar cytoplasm. Isolated cytoplasmic spherules, many of which contain organelles, are found adjacent to the cell wall. (From Wodzicki & Humphreys, *J. Cell Biol.* **56,** 1973. Reproduced by courtesy of the authors.)

Vesicular membrane-bound bodies, associated with cell wall formation during the differentiation of secondary xylem elements in *Pinus strobus,* appeared to originate from the Golgi apparatus (Murmanis & Sachs, 1973). The contents of these bodies apparently were released to the cell wall by fusion of the vesicular membranes of the bodies with the plasmalemma. These bodies were observed during the formation of the cell plate, primary wall and secondary wall. The fibrillar material contained within some of the vesicles structurally resembled the fibrillar component of the existing cell wall. Murmanis & Sachs (1973) assumed that these Golgi-derived vesicles were transporting hemicellulosic, cellulosic and pectic components involved in wall biosynthesis. Although this agrees with the previous results of Brown *et al.* (1970) involving the demonstration of cellulose and pectin in the Golgi apparatus of a Chrysophaceae alga, there is no experimental evidence to indicate the identity of the dark material and the fibrillar material observed by Murmanis & Sachs (1973) within these bodies.

Some interesting cytological differences have been noted between the storage parenchyma and vessel-associated parenchyma cells observed in the secondary xylem of *Robinia, Acer* and *Oeillet* (Czaninski, 1968, 1970, 1972*a*). In mature vessel-associated cells a new wall layer was formed following maturation of the cells. This newly synthesized layer was distinct from the secondary wall and it was formed only on the cell side abutting a vessel element (Czaninski, 1973).

The final maturation stage of cytodifferentiation in the secondary xylem of *Pinus* has been examined with transmission and scanning electron microscopy (Wodzicki & Humphreys, 1972, 1973; Fig. 28). The cytoplasm consisted of an outer layer, lining the cell wall, and a web-like intravacuolar cytoplasm enclosed by vacuolar membranes and immersed in vacuolar sap. The formation of cytoplasmic spherules, released to the vacuole, and the breakdown of the intravacuolar cytoplasm, signaled the beginning of autolysis (Wodzicki & Humphreys, 1973). The autolytic process may be analogous to the intravacuolar breakdown observed in senescing cells (Matile & Winkenbach, 1971).

Wound vessel members

In addition to primary xylem and secondary xylem, a third type of xylem element arises either as a result of rupturing a vascular bundle or by hormonal treatment of excised plant tissues. Most of the ultrastructural research on wound vessel members, resulting from wounding stem internodes of *Coleus,* has been conducted by Hepler and his co-workers (Hepler & Newcomb, 1963, 1964; Hepler *et al.,* 1970; Hepler & Fosket, 1971; Hepler *et al.,* 1972).

Several facets of the ontogeny of wound vessel members of *Coleus* have been examined. The reported banding of the cytoplasm (see p. 71) and clustering of cytoplasmic organelles in correspondence with secondary wall thickenings (Hepler & Newcomb, 1963) was not verified in a subsequent investigation (Hepler *et al.,* 1970). Cytoplasmic microtubules, however, were localized specifically over the secondary wall thickenings (Hepler & Newcomb, 1964; Hepler *et al.*, 1970; Hepler & Fosket, 1971), and similar observations have been made on differentiating cells of primary xylem (Maitra & De, 1971; Srivastava & Singh, 1972) and secondary xylem (Robards & Kidwai, 1972). As discussed earlier in this chapter, Hepler & Fosket (1971) made a careful study of the role of microtubules in vessel member cytodifferentiation in *Coleus*. Studies have also been conducted on the lignification (Hepler *et al.*, 1970) and cytochemical localization of peroxidase activity in wound vessels of *Coleus* (Hepler *et al.*, 1972), subjects which will be covered later in this chapter.

Cronshaw (1967) has examined the ultrastructure of differentiating tracheary elements induced to form in tobacco pith cultures in the presence of IAA and kinetin. Unfortunately, no observations were made on the early phases of cytodifferentiation prior to the deposition of the secondary wall. The microtubules appeared to be attached to the plasma membrane, with the direction of orientation of the microtubules mirroring that of the most recently deposited cellulosic fibrils. Membranous inclusions were observed between the plasma membrane and the developing wall thickenings, as previously noticed in *Acer rubrum* (Cronshaw, 1965*a,b*) in differentiating fibers that were developing secondary walls. Invaginations of the plasma membrane contained loosely organized fibrillar material, and these invaginations were situated in regions which had not, as yet, exhibited secondary wall thickenings. However, dictyosome-derived vesicles apparently were not associated with secondary wall depositions (Cronshaw, 1967).

Wright & Bowles (1974) examined the effects of auxin and zeatin on the polysaccharides deposited in the cell walls during cytodifferentiation in explants of lettuce pith parenchyma. A radioactive precursor was employed to follow polysaccharide biosynthesis within isolated Golgi and ER fractions. Zeatin-induced cytodifferentiation was accompanied by nearly a twelvefold increase in the amount of radioactivity in polysaccharides, and the distribution of label among the different sugars was affected by the presence of zeatin (Wright & Bowles, 1974).

Lignification

Lignin is an aromatic polymer of *p*-hydroxyphenylpropanoid subunits which impregnates the intercellular layer, primary and secondary walls of tracheary elements, fibers, and sclereids. Initially erythrose-4-phosphate joins with phosphoenolpyruvate to form deoxyarabinoheptulosonic acid phosphate; the pathway follows successively shikimic acid, chorismic acid and prephenic acid to the phenylalanine and tyrosine pools (Berlyn, 1970). The action of the enzyme L-phenylalanine ammonia-lyase (PAL) results in the conversion of phenylalanine to cinnamic acid, whereas L-tyrosine ammonia-lyase produces *p*-coumaric acid from tyrosine (Neish, 1964). The cinnamyl alcohols, oxidized by peroxidases and other oxidative enzymes, lead to the polymerization and co-polymerization of coniferyl, sinapyl and *p*-coumaryl alcohols (Brown, 1966). Glucosides, such as coniferin and syringin, may furnish lignin precursors in certain plant groups (Freudenberg, 1964). The lignin molecule is formed chiefly by dehydrogenative polymerization of 1-(4-hydroxyphenyl) allyl alcohol, i.e., *p*-hydroxycinnamic alcohol, 1-(4-HO-C_6H_4)-CH = $CHCH_2OH$ and, especially, its phenyl-substituted mono- and di-methoxylated derivatives (the so-called phenylpropane or C_6-C_3 series of precursors; Stewart, 1969). Lignification, as related to the cytodifferentiation of tracheary elements, has been reviewed by Berlyn (1970), Roberts (1969) and Stewart (1969). Further information on the biochemistry of lignin formation is available in reviews by Freudenberg (1964, 1965), Neish (1964, 1965), Kratzl (1965), Schubert (1965), Isherwood (1965), Brown (1966, 1969) and Sarkanen & Ludwig (1971).

According to Hepler *et al.* (1970), potassium permanganate specifically stains lignin, and this technique was employed to follow the pattern of lignification in differentiating wound vessel members in *Coleus* (Fig. 29). Lignin was initially detected in the middle lamella and primary wall directly beneath the bands of secondary wall thickening; as wall thickening developed, the lignification became progressively more extensive. Identical results were obtained when the developing wound vessel members were examined with ultraviolet light fluorescence (Hepler *et al.*, 1970). This confirms the observations of Wardrop & Bland (1959); lignin appeared initially in the middle lamella and primary wall in the corners of immature tracheids of *Pinus radiata* when viewed with ultraviolet light microscopy. A question has been raised concerning the use of potassium permanganate as a specific cytochemical reagent for lignin localization (O'Brien, 1974). Hepler and co-workers (1970) cited three reasons for assuming that potassium permanganate specifically combines with lignin. The fixative stains only the primary wall at sites where the secondary wall is attached and presumably is not combining with cellulose, hemicellulose, pectin or protein. The pattern of potassium permanganate staining is identical with lignin fluorescence resulting from ultraviolet irradiation. Since potassium permanganate is an oxidizing agent, the electron opacity is probably due to a complex between lignin and manganese dioxide (Hepler *et al.*, 1970). The Mäule reaction, studied by Crocker (1921), involves the reduction of potassium permanganate to manganese dioxide in the presence of lignin. O'Brien (1974) stated the electron opacity after fixation by potassium permanganate cannot be considered a reliable indicator of lignin, since starch, mucilage and vacuolar deposits may all be electron opaque after fixation by potassium permanganate (O'Brien, 1972).

A specific organelle or structure has not, as yet, been identified with the process of lignification. Experiments with labeled phenylalanine (Wooding, 1968) and cinnamic acid (Pickett-Heaps, 1968) indicated that the secondary thickenings were labeled, but that no specific organelles were involved. Hepler *et al.* (1972) found a similarity between the cytochemical localization of peroxidase and lignin in *Coleus* wound vessel members; the enzymes involved in lignin

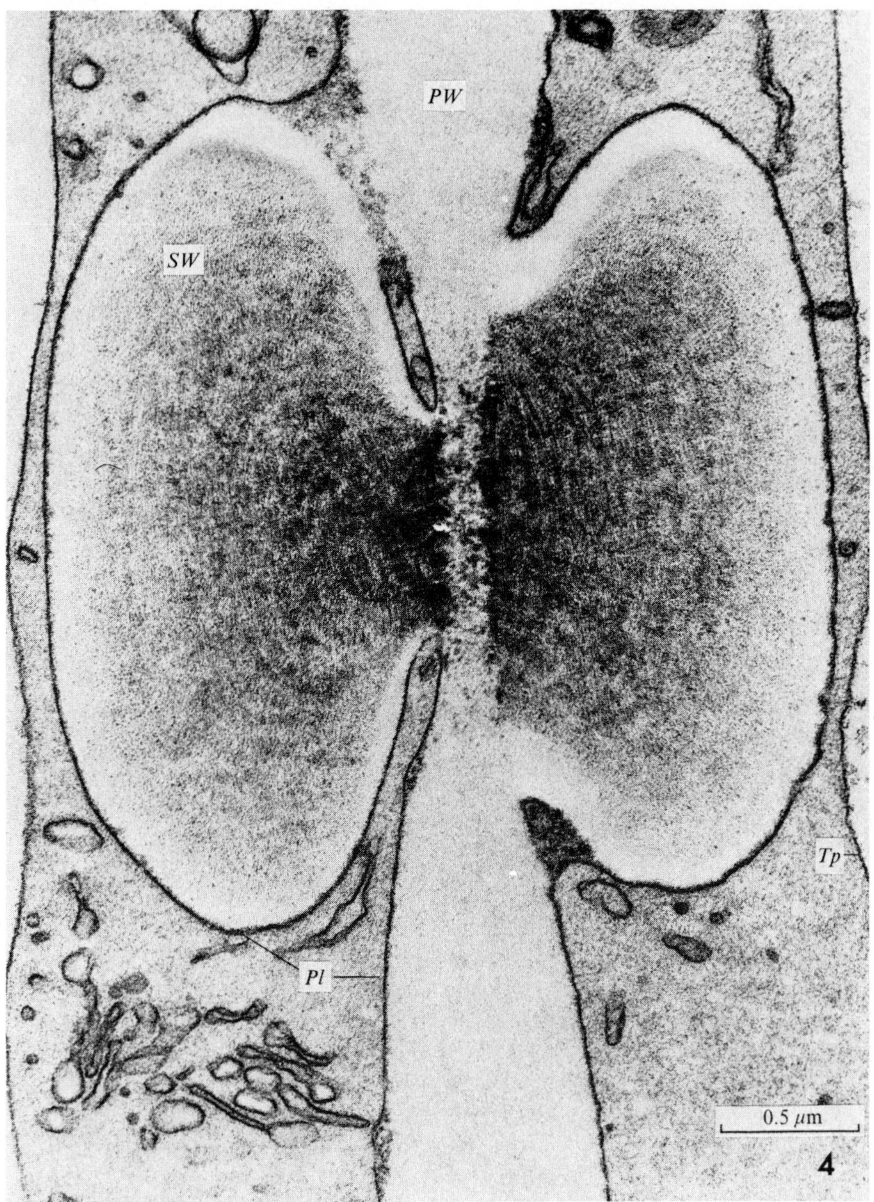

Fig. 29. Fine structure of lignin deposition in developing secondary walls (*SW*) during cytodifferentiation of wound vessel members of *Coleus*. Note the narrow constriction of the primary wall (*PW*) where the ridges of secondary wall are formed. Potassium permanganate was employed both as a fixative and as a cytochemical reagent for the lignin component of the cell wall. Lignin deposition is highest at the junction of the primary and secondary walls, and the fibrillar nature of the electron-opaque material in the secondary wall is evident in this electron micrograph. *Pl*, plasmalemma; *Tp*, tonoplast. (From Hepler, Fosket & Newcomb, *Am. J. Bot.* **57,** 1970.)

biosynthesis are probably localized within the regions of the wall exhibiting lignification (Hepler *et al.*, 1970).

Although the following remarks do not concern ultrastructural cytology, the findings are pertinent to our discussion of the phenomenon of lignification. Rubery & Fosket (1969) observed that changes in PAL activity in cultured *Coleus* stem segments and soybean callus cultures followed the same time course as the formation of lignified xylem elements. The experiments, however, were not performed to show whether the PAL activity resulted from enzyme synthesis *de novo* or the activation of an existing system. Obroucheva (1969) indicated that protein synthesis was not necessary for the initiation of lignification. Although mitotic activity and growth were strongly inhibited following treatment of young maize roots with X-ray irradiation (10 000 r) and chloramphenicol, respectively, the rate and character of lignification was completely unaffected by these treatments. The effects of the same treatments on other aspects of xylem differentiation were not mentioned (Obroucheva, 1969). Fosket & Miksche (1966) found that X-ray irradiation (4000 r) reduced xylogenesis by fifty per cent in *Coleus* slices and strongly blocked protein synthesis. Berlyn and co-workers (1969, 1970) reported that five genes were involved in the pre-chorismic acid portion of the aromatization pathway. Peroxidases may be involved in the terminal stage of lignification (Hepler *et al.*, 1972).

Although decisive experiments have not been performed, it appears unlikely that the phenomenon of lignification *per se* has any direct bearing on the early events occurring during the initiation of cytodifferentiation.

Cytochemical evidence

The localization of substances by chemical techniques during cytodifferentiation is subject to interpretation. First, the specificity and sensitivity of the technique employed and the absence of diffusion or fixation artifacts must be considered. Second, it must be established that the localized substance, at a given site, is a critical variable in some developmental stage of cytodifferentiation. Some localized substances may be related to the basal metabolism of the cell or may have functional roles unrelated to cytodifferentiation. Complications arise with cultured plant tissues because of the induction of enzymatic activity by wounding at the time of excision. Enzymes with increased activity arising from tissue damage included peroxidases, acid phosphatase, DNase, thymidine monophosphatase and invertase (Yeoman & Aitchison, 1973). The in-vitro cultural conditions imposed on the tissue system may stimulate metabolic pathways that have no parallel or developmental significance in the intact plant.

The localization of IAA in tracheary elements during the middle and late stages of cytodifferentiation has been attempted by microradioautography. Similar experiments with labeled synthetic auxins have not been performed, nor have experiments with gibberellins or cytokinins. The rationale for such an experiment at the initiation of cytodifferentiation appears uncertain. Hormones probably have rapid transient effects, whereas the localization of a label with microradioautography involves the incorporation of the active compound containing the label. The assimilation of the label may involve not the whole molecule, but some metabolized fragment of the original tagged molecule unrelated to the hormonal effect. The association of labeled tritiated IAA (Sabnis, Hirshberg & Jacobs, 1969) and [^{14}C]IAA (Wangermann, 1970; Gee, 1972) with secondary wall thickenings has been reported. Gee (1972) treated wounded *Coleus* internodes by implanting anion exchange resin beads loaded with methyl-labeled [^{14}C]IAA; he found no evidence that prospective wound xylem cells contained more IAA than neighboring cells at any stage of their induced cytodifferentiation.

Phosphatase activity may be involved in cytodifferentiation. Gahan & Maple (1966) observed a change from a particulate to a diffuse distribution of β-glycerophosphatase during protoxylem differentiation in the roots of *Vicia faba*; they suggested that cell autolysis occurred following the release of hydrolytic enzymes from lysosomes. The change in the distribution of acid phosphatase activity was accompanied by a marked decrease in the time of incubation required to exhibit the enzyme activity from twenty minutes for meristematic cells to two-to-four minutes for cells at a later stage of cytodifferentiation (Gahan & Maple, 1966). Sexton & Sutcliffe (1969) observed that β-glycerophosphatase activity was confined to particulate sites after secondary wall thickening had occurred in differentiating xylem elements of *Pisum,* but that diffuse localization coupled with short incubation times was associated with autolysis of the protoplast. Strong acid phosphatase activity was associated with the cytodifferentiation of water-conducting elements in several bryophytes (Hébant, 1973) and ferns (Shaykh & Roberts, 1974). During xylem differentiation in wounded *Coleus* stem internodes, the pattern of acid phosphatase activity coincided with the formation of pits in the walls of the tracheary elements (Jones & Villiers, 1972). Since enzyme activity was initially observed about two weeks after wounding, it is difficult to understand why acid phosphatase activity was *absent* during the first week after wounding during a period of intense xylem differentiation. The rise in activity was probably associated with the establishment of a functional role for the newly formed vascular system. Phosphatase activity associated with the contact cells of wood rays in *Acer saccharum* was concentrated on the large pits between the contact cells and the vessels, and the activity of the enzyme was associated with the release of sucrose into the xylem sap (Sauter, 1972; Sauter, Iten & Zimmermann, 1973). Acid phosphatase activity observed in the parenchyma cells of the secondary vascular tissues of sycamore and locust trees during the spring was absent in the autumn and winter (Catesson & Czaninski, 1968). High phosphatase activity in the differentiating secondary xylem of beech (*Fagus sylvatica* L.) utilized ATP or ADP as substrates in the nucleoli, nucleoplasm, nuclear envelope, ER, plastids, mitochondria, and at the surface of the plasmalemma (Robards & Kidwai, 1969*a*).

The capability of various plant hormones, including ethylene, to induce either the activation or synthesis of peroxidase isoenzymes seems well established (Lavee & Galston, 1968; Ritzert & Turin, 1970; Ridge & Osborne, 1970; Lee, 1972; Gaspar, Khan & Fries, 1973). There is a possibility that one or more peroxidases may be involved during the initiation of cytodifferentiation (Borchert, 1974). Verma & van Huystee (1970) studied the development of peroxidase isoenzymes in cell clumps of different sizes obtained from a suspension culture of peanut (*Arachis hypogaea* L.) cotyledon. A new isoenzyme appeared in cell masses of 500 μm in diameter during the initiation of cytodifferentiation. As tracheary element differentiation proceeded, in cell masses 2–4 mm in diameter, a second peroxidase isoenzyme appeared. Changes in soluble and wall-bound peroxidase activity were observed during xylogenesis in cultured Jerusalem artichoke tissues, but no evidence was given to indicate any direct relationship between the enhanced enzymatic activity and cytodifferentiation (Minocha & Halperin, 1973). The strongest evidence that peroxidase plays a role in the initiation of cytodifferentiation involves recent findings concerning the effects of ethylene on the process (see Chapter 3). Ethylene (0.1 ppm) increased wall-bound peroxidases and wall-associated hydroxyproline-rich proteins in pea epicotyl tissues (Ridge & Osborne, 1970). Evidence suggests that ethylene is associated with the reorientation of cellulosic fibrils during cell wall growth (Ridge, 1973) and that this effect involves the cross-linkages by glucosidic bonds of hydroxyproline-rich proteins with wall polysaccharides (Osborne *et al.*, 1972). Exogenous proline, in the presence of kinetin, was observed to stimulate peroxidase activity (Parish, 1968). Although the incorporation of proline in an auxin–cytokinin medium enhanced xylogenesis in internodal segments of *Coleus* (Roberts & Baba, 1968*b*), a similar response was not observed in explants of lettuce pith (Roberts, unpublished observations). The significance of the observation that ethylene and IAA have dissimilar effects on cell wall deposition is not known (Sargent *et al.*, 1973). Possibly one of the early effects of auxin on the initiation of cytodifferentiation is to stimulate the formation of ethylene at some stage of the cell cycle.

An ultrastructural study of peroxidase localization in wound vessel members of *Coleus* indicated cytochemical staining in the reticulate secondary wall and in the primary wall where the secondary wall thickenings occurred (Hepler *et al.*, 1972; Fig. 30). The electron-opaque localization was achieved by incubation of the specimens in 3,3′-diaminobenzidine (DAB) and hydrogen peroxide. In the cytoplasm of the differentiating tracheary elements, staining was observed in the plasmalemma, particularly at the sites overlaying the secondary wall thickenings, and in dictyosomes and dictyosome vesicles. Peroxidase localization in the plasmalemma and in dictyosome vesicles was also observed by Innocenti (1973)

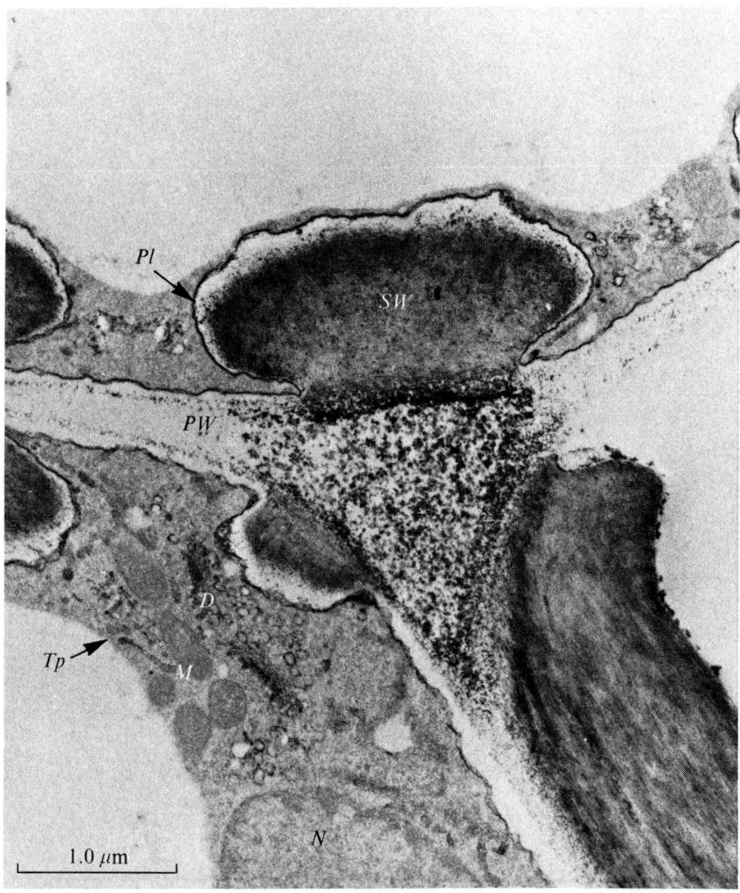

Fig. 30. Cytochemical localization of peroxidase activity during cytodifferentiation of wound vessel members of *Coleus*. After culture of stem internode segments on a xylogenic medium, specimens of differentiating tissue were fixed in glutaraldehyde, incubated in diaminobenzidine and hydrogen peroxide and postfixed in osmium tetroxide. The secondary wall (*SW*) and the regions of the primary wall (*PW*) associated with the secondary wall are heavily stained. Staining is more intense toward the periphery of the secondary thickenings than it is in the core. High peroxidase activity is evident at the junction of primary and secondary walls. The plasmalemma (*Pl*) and dictyosome vesicles (*D*) are stained, whereas the tonoplast (*Tp*), nucleus (*N*), and mitochondria (*M*) remain unstained. (From Hepler, Rice & Terranova. Reproduced by permission of the National Research Council of Canada from *Can. J. Bot.* **50,** No. 5, 1972, pp. 977–83.)

in differentiating metaxylem elements in *Allium* root tips. According to Hepler *et al.* (1970), the similarity of localization of peroxidase and lignin supports the concept that this enzyme functions in lignification. Harkin & Obst (1973) reported the presence of peroxidase activity in lignifying cells of certain trees and they expressed the view that peroxidase was the only enzyme present in these tissues capable of the polymerization of *p*-coumaryl alcohol to lignin. In addition to the ultrastructural study of peroxidase by Hepler *et al.* (1972), the close association of peroxidase activity with lignification has been shown in other cytochemical studies (Jensen, 1955; Higuchi, 1957; Stafford, 1962; Lipetz & Garro, 1965; Gagnon, 1968: see reviews by Van Fleet, 1962; Roberts, 1969).

On the other hand, several reports have indicated that the cytochemical localization of lignin and peroxidase, respectively, did not coincide in differentiating xylem cells. Although immature tracheary elements exhibited a positive reaction for peroxidase activity, the enzyme apparently disappeared at the onset of lignification (Wardrop & Bland, 1959; Koblitz, 1961). Peroxidase activity could not be detected in lignified cells of the primary xylem in root tips of *Allium* (De Jong, 1967) and in the secondary xylem of *Robinia* and *Acer* (Czaninski & Catesson 1969). Procedural artifacts are probably not involved because Hepler *et al.* (1972), Innocenti (1973) and Czaninski & Catesson (1969) employed essentially the same procedure, i.e., the diaminobenzidine reaction coupled with electron microscopy. The pH of the incubation medium may be a factor to take into consideration (see discussion by Innocenti, 1973).

Aside from the unresolved question of peroxidase activity and the lignification of tracheary elements, it appears that the more intriguing question is whether or not peroxidases are a critical variable in the auxin-induced initiation of cytodifferentiation. Certainly more research is needed on the roles of peroxidases during cytodifferentiation.

7

Nutritional factors

Nearly all research on the relation of nutrition to xylem differentiation has been concerned with the relative effectiveness of various carbohydrates in initiating the CDS. As far as we know, various minerals, vitamins, amino acids and other nutritive supplements should be considered as controlling but not critical variables in the initiation and normal development of tracheary elements. These nutritive factors doubtless play important roles in the basal metabolism necessary for normal cell growth and division, but they probably are not involved in specific and unique roles in cytodifferentiation. Suggestions have been made involving certain minerals in cytodifferentiation (Duffy, 1965; Swain & Rier, 1968; Rier, 1970), but these interpretations are open to question. The CDS cannot be examined, as a separate process, in a cell population with an abnormal basal metabolism stemming from mineral deficiency. Unfortunately, most research workers have been unaware of the remarkable array of mineral elements and other substances present as contaminants, in relatively high concentrations, in commercial agar (Pierik, 1971).

We are faced with a complex problem when we consider carbohydrates functioning either as controlling or critical variables in the CDS. Carbohydrates are necessary in living cells as a carbon source for energy demands, nucleotide synthesis, cell plate and primary wall formation, plus a host of other metabolic chores that constitute roles as controlling variables. Are carbohydrates to be considered as critical variables? Clearly certain saccharide monomers must be present during the deposition of the secondary wall in the differentiation of tracheary elements; thus carbohydrates should be considered as critical variables at that period of cytodifferentiation. At what prior time, before cell wall deposition, these monomers become critical variables in the CDS is not known. The suggestion has been made that α-glucosyl disaccharides may be instrumental in inducing cytodifferentiation (Jeffs & Northcote, 1967), perhaps at the point of determination, but this is a debatable issue. The action of sucrose may be analogous to the effect of lactose on the induction of β-galactosidase synthesis in certain strains of *Escherichia coli* (Jacob & Monod, 1963).

Xylogenesis in cultured plant tissues requires the presence of an exogenous source of carbohydrates (Wetmore & Rier, 1963; Fosket & Roberts, 1964). Although the minimal amount of exogenous carbohydrate necessary to sustain cytodifferentiation probably varies from one tissue system to another, Jeffs &

Northcote (1967) calculated that a value of 0.75 per cent sucrose was present at the site of xylogenesis in bean callus. Wright & Northcote (1972) critically examined the carbohydrate requirement for xylogenesis in sycamore callus. Although a callus may achieve moderate growth with a sugar that is inefficiently metabolized, cytodifferentiation may not occur because of an inadequacy of some intermediate metabolite required for cell division and cytodifferentiation. Since tracheary element formation must proceed to a relatively late stage for visual recognition with the light microscope, it is possible that the initiation of cytodifferentiation occurs, but is aborted because the subsequent metabolic demands cannot be met (Wright & Northcote, 1972). These workers found that sucrose, glucose and fructose were invariably formed by interconversions from *any* exogenous sugar which supported the growth of sycamore callus. Sucrose can be metabolized by at least four different enzyme systems: sucrose phosphate synthetase (E.C. 2.4.1.14), sucrose synthetase (E.C. 2.4.1.13) and two different invertases (β-D-fructofuranoside fructohydrolase, E.C. 3.2.1.26) which occur with acid and alkaline pH optima (Ricardo & ap Rees, 1970). Invertase activity may also be regulated by tissue levels of sucrose, glucose and fructose (Kaufman *et al.*, 1973). Since the relative amounts of these enzymes apparently vary in different tissues and within different parts of the same tissue (Lyne & ap Rees, 1971; Hawker, 1971), higher plants possess an intricate system of controls over the cellular concentration of sucrose (Wright & Northcote, 1972). Attempts to determine optimal levels of various saccharides for cytodifferentiation are probably without justification because of the complexity of the regulatory systems involved in carbohydrate interconversions.

Some experimental results obtained from cultured plant tissues are difficult to interpret because of the use of autoclaved sugars in the culture media. Ball (1953) demonstrated that autoclaving a 3 per cent sucrose medium resulted in a medium containing 0.7 to 0.9 per cent of a mixture of D-glucose and D-fructose in addition to sucrose. Callus of *Sequoia* grown on culture media containing either autoclaved or filter-sterilized sucrose produced two distinctly different types of callus growth; callus growth on the autoclaved medium was enhanced compared to that on the filter-sterilized medium (Ball, 1953). Romberger & Tabor (1971) found that autoclaving sucrose, in the presence of agar, improved the growth of shoot apical meristem cultures. They reported that an 'agar inhibitory effect' was reversed by nearly half following the substitution of D-glucose and D-fructose for sucrose. Apparently some results on cytodifferentiation may be influenced by reactions between sugar degradation products and agar contaminants. A different view was given by Stehsel & Caplin (1969) following an investigation of the growth of carrot tissue in media containing various levels of sucrose, D-glucose, or D-fructose, sterilized by either autoclave or filtration. Filter sterilization produced the greatest growth for each sugar. Severe growth inhibition was produced by autoclaved D-glucose, while autoclaved D-fructose gave complete growth inhibition at two per cent concentration. The effect of the treatments on cytodifferentiation was not reported (Stehsel & Caplin, 1969). Wright & Northcote (1972) found that autoclaved D-fructose produced less growth and cytodifferentiation in sycamore callus than when the sugar was filter-sterilized. In explants of lettuce pith, Cawthon (1972) found considerably fewer tracheary elements in the presence of three per cent D-fructose compared with either three per cent D-glucose or three per cent sucrose when all of these sugars were filter-sterilized. The decomposition of D-fructose occurs at room temperature; when D-fructose accumulates in the explant tissue and in the presence of certain minerals or other compounds, its degradation is probably facilitated (G. P. Rédei, personal communication). The toxicity of autoclaved D-fructose stems mainly from a suppression of the activity of fructose-1, 6-diphosphate aldolase by the degradation products; this, in turn, causes a shortage of metabolites in subsequent reactions (Rédei, 1974). Sugar degradation products may have an inhibitory effect on xylogenesis, whereas the enhancement of responses following autoclaving could involve an alteration of some agar contaminant. The degradation of sugars also involves reactions with amino acids present in the medium (Peer, 1971).

An ultrastructural study of galactose toxicity indicated that this sugar probably affects some factor involved in the maintenance of membrane permeability (Ernst, Arditti & Healey, 1971). Relatively little xylem formation was observed when this sugar was the carbon source in media supporting callus growth (Wright & Northcote, 1972; Jeffs & Northcote, 1967).

Some lines of evidence suggest that sucrose and certain other sugars may possess unique capabilities for the initiation of cytodifferentiation in particular circumstances. In the callus study of Wetmore & Rier (1963), the anatomy of the vascularized nodules was completely transformed by increasing the concentration of the exogenous sucrose at a fixed auxin concentration. In one experiment with sycamore callus, exogenous raffinose apparently induced vascular nodule formation in the absence of exogenous cytokinin (Wright & Northcote, 1972), but this has not been observed in other tissue systems (Ball, 1955; Jeffs & Northcote, 1967). Torrey *et al.* (1971) suggested that sucrose may be contaminated with traces of naturally occurring plant hormones such as cytokinins; this possibility should be investigated. Sucrose displays some unusual sequential 'hormone-like' effects with auxin. Callus implanted initially with sucrose formed only tracheary elements; callus treated initially with IAA, followed by sucrose, formed vascular nodules containing xylem, phloem and a cambium (Jeffs & Northcote, 1967). The interpretation offered was that auxin initiates mitosis and the recently divided cells were somehow stimulated into differentiation with the addition of sucrose; the opposite order of treatment was ineffective because the cells had not divided (Jeffs & Northcote, 1967). Gametophytes cultured on one or two per cent sucrose exhibited a lag period of ten days before the formation of tracheary elements was observed, whereas cytodifferentiation occurred immediately in gametophytes grown on five per cent sucrose–agar (DeMaggio, 1972). In addition, DeMaggio (1972) found that exogenous sucrose was highly effective in initiating cytodifferentiation in fern gametophytes in the absence of any other xylogenic supplements. Possibly the gametophytes were capable of synthesizing xylogenic levels of hormones, but endogenous carbohydrates were limiting cytodifferentiation. The pretreatment of explants of lettuce pith with sucrose prior to culturing them on a xylogenic medium apparently shortened the lag phase of xylem differentiation, suggesting that some product of sucrose metabolism may stimulate the hormonal initiation of xylogenesis (Cawthon, 1972).

Can the cytodifferentiation responses shown by sucrose be completely replaced by an equivalent concentration of a combination of D-glucose and D-fructose, or does the disaccharide molecule possess some unique morphogenetic ability (Jeffs & Northcote, 1967)? Bean callus was examined following treatment with implanted 'induction wedges' of agar containing IAA and various sugars (Jeffs & Northcote, 1967). Control callus, treated with wedges of agar alone, never produced tracheary elements. Wedges containing a disaccharide plus IAA induced organized xylem and vascular nodules, whereas a combination of D-glucose and D-fructose with IAA produced only a trace of xylem and no vascular nodules. Sucrose, trehalose and maltose, respectively, gave similar results; the conclusion was drawn that α-glucosyl disaccharides possess some unique morphogenetic property. It is important to point out that the cultural conditions for tracheary element differentiation and vascular nodule formation are *completely* different. Cytodifferentiation occurs within four days in pith explants of lettuce (Dalessandro & Roberts, 1971), whereas vascular nodule formation requires between thirty to sixty days after implantation of the inductive wedge. The callus employed in the experiments of Jeffs & Northcote (1967) also received xylogenic growth regulators from both the inductive wedge and the maintenance medium on which the callus was resting. Since the media were autoclaved, the poor response of the monosaccharide mixture probably resulted from D-fructose degradation. Nevertheless, in pith explants of lettuce a quantitative comparison of xylogenesis in the presence of filter-sterilized sucrose with equivalent amounts of combinations of filter-sterilized D-glucose and D-fructose revealed that greater numbers of tracheary elements were consistently formed with sucrose than with the combination of the two monosaccharides (Cawthon, 1972). A possible explanation is that sucrose can be degraded *in vivo* by sucrose synthetase (E.C. 2.4.1.13) to form NDP-glucose (Delmer & Albersheim, 1970), which can then directly enter the reactions producing primary and secondary wall monomers. On the other hand, the assimilation of either D-glucose or D-fructose requires the expenditure of ATP energy for phosphorylation, and these sugar phosphates would be more readily available as substrates for the glycolytic pathway than for wall metabolism.

Minocha & Halperin (1974) examined the effects of various carbohydrates and carbohydrate derivatives on auxin-induced xylogenesis in Jerusalem artichoke explants. The carbohydrates were autoclaved with the other components of the medium. Of the various substances tested as possible carbohydrate sources for cytodifferentiation, sucrose, glucose and trehalose were equally effective in the initiation of xylogenesis. However, soluble starch (4 per cent) was the most effective carbohydrate examined, and thirty to forty per cent of the total number of cells formed tracheary elements after fourteen days incubation in the presence of NAA (1 mg/l) and benzyladenine (5 mg/l). Since soluble starch is a natural plant product, the stimulatory effect might have resulted from traces of some growth regulator in the reagent. The combination of sucrose (2 per cent) and glucose (2 to 4 per cent) inhibited cytodifferentiation eighty to ninety per cent, and the effect was apparently non-osmotic. We may speculate that this inhibition involved sugar nucleotide formation. Sucrose degradation produces NDP-glucose, by the action of sucrose synthetase, and we may assume that this nucleotide monomer is required for wall formation during cytodifferentiation. Since this reaction is freely reversible (Delmer & Albersheim, 1970), the formation of NDP-glucose was probably suppressed by the large amounts of exogenous glucose employed by Minocha & Halperin (1974).

During cytodifferentiation, at some period after cell division, cell enlargement requiring primary wall formation generally occurs. At some unspecified time thereafter, primary wall development diminishes and secondary wall formation increases. The biosynthesis of the primary wall depends upon the formation of UDP-D-galactose, UDP-D-galacturonic acid, UDP-D-arabinose and UDP-L-rhamnose; the synthesis of each of these nucleotide sugars requires the activity of a specific epimerase (Northcote, 1968, 1969*a*). As the cell CDS progresses, a gradual shift occurs from the biosynthetic pathways leading to monomers for primary wall formation to an accumulation of secondary wall monomers. The latter compounds include UDP-D-glucose, UDP-D-glucuronic acid and UDP-D-xylose (Fig. 31). The alteration in the type of polysaccharide synthesized by the cell can be effected by influencing any part of the carbohydrate biosynthesis, from interconversion of the nucleoside diphosphate sugar by epimerase enzymes, to an action on the transglycosylases or by changing the availability of the substrate molecules (Northcote, 1971). The xylose : arabinose ratio was employed by Jeffs & Northcote (1966) to express quantitatively the relative degree of cytodifferentiation, i.e., secondary wall formation, occurring in bean callus and sycamore stem tissues.

One question currently under investigation is whether or not these enzymes and nucleotide sugars can be detected in actively differentiating xylem tissues. Rubery (1972) demonstrated that UDP-D-glucose: NAD oxidoreductase, involved in the conversion of UDP-D-glucose to UDP-D-glucuronic acid, exhibited greater activity in differentiating xylem than in the cambium isolated from sycamore trees. Subsequently, Rubery (1973) found that UDP-D-glucose 4-epimerase, involved in converting UDP-D-glucose to UDP-D-galactose, was present in sycamore cambium and in differentiating xylem; the enzyme appeared to be localized in the non-lignifying primary-walled cells common to both fractions. Previously, Boothby (1972) had reported that extracts of sycamore cambium were capable of the conversions of UDP-D-glucose to UDP-D-glucuronic

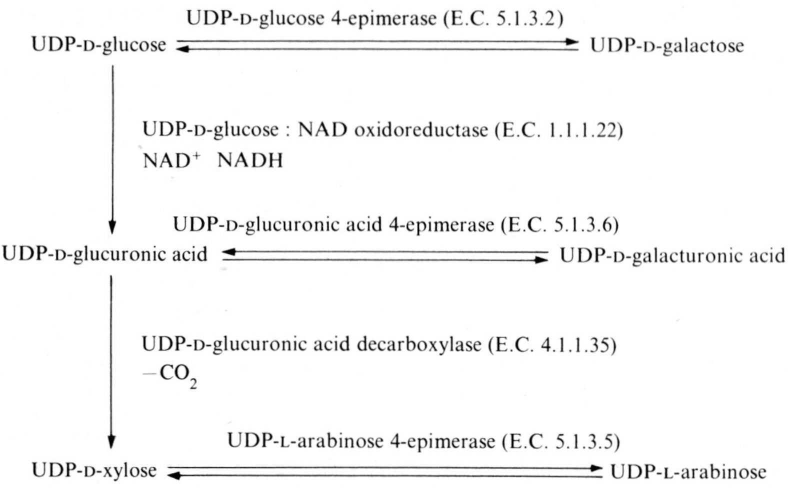

Fig. 31. Enzymatic interconversions between carbohydrate monomers involved in the biosynthesis of the cell wall. The relative enzymatic activity of the epimerases may be an important factor in shifting the metabolism from the production of primary wall monomers to secondary wall monomers during cytodifferentiation. (Adapted from Northcote, 1968.)

acid, and the latter into UDP-D-xylose and UDP-L-rhamnose, but that none of the corresponding galactose series of monosaccharides were formed. Although this finding would support the hypothesis of Northcote (1969*a*,*b*; 1971) concerning the lack of epimerase activity during secondary wall formation, it has not been verified. In fact, Cumming (1970) identified UDP-D-galactose and UDP-D-arabinose in cambial and young xylem tissue of *Larix*.

The possibility that certain cytodifferentiation responses arising from the incorporation of carbohydrates in culture media may, to some extent, stem from osmotic effects will be examined in Chapter 8.

In resumé, a clearer understanding of the roles of carbohydrates in cytodifferentiation is seriously needed. Sucrose is the most efficient carbon source for stimulating cytodifferentiation, and several experiments have demonstrated the unique qualities of this sugar. Its unusual xylogenic ability may stem from some unknown contaminant (Torrey *et al.*, 1971), molecular configuration (Jeffs & Northcote, 1967), facility in the formation of monomers for wall synthesis, or an unknown reason. It is premature to speculate on the possible inductive influence of nucleotide diphosphate sugars of the glucose series on the initiation of xylogenesis or of the activity or inactivity of the various epimerases on xylem differentiation.

8

Environmental influences

In this chapter we will consider some external controlling variables capable of modifying either the initiation or progress of cytodifferentiation; these factors should not be considered as critical variables in xylem differentiation. This short discussion is not intended to serve as an ecological review of xylem differentiation, since the literature examined has been restricted to a relatively few studies.

Water

Water stress has been studied primarily from the standpoint of the cell receiving too little water rather than too much. Plant cells in suspension cultures thrive under conditions of immersion in a liquid medium, provided that the cultures receive some agitation. The movement of the medium serves to keep the cells and cell aggregates evenly distributed and facilitates sufficient gaseous exchange between the culture medium and the atmosphere within the vessel (Street, 1974).

Water tupelo (*Nyssa aquatica* L.) and green ash (*Fraxinus pennsylvania* Marsh.) are two interesting examples of the adaptation of certain trees to wet habitats. These hydrophytes typically grow under conditions of poor aeration of the transpiration stream. Such plants have evolved an alternative means for gas exchange in the region of the differentiating vascular tissues. Hook & Brown (1972) reported that the amount of tension necessary to move air across the vascular cambium of water tupelo and green ash was relatively low, suggesting a free gaseous exchange with the atmosphere. Both of these trees were found to possess prominent intercellular spaces among the cambial ray initials. These spaces formed an interconnecting system between the xylem and phloem ray cells, and similar openings in the cambium were absent from several species of mesophytic trees examined by Hook & Brown (1972). This unusual anatomical adaptation is undoubtedly important in the regulation of gaseous exchange processes required for the cytodifferentiation processes of the secondary vascular tissues in these plants.

Doley (1970) subjected *Liriodendron* seedlings to simulated drought conditions. He observed that vessel elements, fibers and ray parenchyma cells in the newly formed secondary xylem were narrower in the radial direction, that vascular rays were increased in width, and that more vertical parenchyma cells were formed than under conditions of rapid growth. However, the differences between the anatomy of the secondary xylem formed at various water potentials diminished toward the shoot apex, and these anatomical differences may have resulted from an impairment in the synthesis and translocation of growth factors (Doley, 1970).

Since increasing the concentration of carbohydrates in the nutrient medium lowers the availability of water to cultured tissues, the resulting water stress may influence cytodifferentiation processes. Doley & Leyton (1970) examined the effects of water potential in the presence of various growth regulators, on the development of wound callus in *Fraxinus*. There was a greater production of xylem elements after the water potential of the medium had been lowered with either sucrose or polyethylene glycol. At each water potential value there was an optimum IAA concentration for xylem differentiation, and this optimum value increased as the water potential became more negative. Doley & Leyton (1970) distinguished between the differentiation of sclereids and xylem cells in the callus of *Fraxinus*. Sclereid initials were either daughter cells from the subsurface meristem or cells incapable of division that had expanded following surface rupture. Both expansion and surface suberization produced physical resistance which was related to sclereid formation (Doley & Leyton, 1970). Xylem differentiation in *Fraxinus* callus, however, was invariably associated with a more or less continuous vascular cambium. Doley & Leyton (1970) indicated that it was impossible to separate the direct effects of auxin on xylem differentiation from the effects of pressure and water potential. Kirkham, Gardner & Gerloff (1972) observed that cell division, as measured by DNA increase, was greatly stimulated by an increased turgor pressure between five and six bars, whereas cell enlargement was stimulated as turgor increased above three bars. There is a possibility that water potential could influence xylogenesis indirectly by affecting the rate of cell division. Water stress may have cytological effects on the Golgi apparatus (Schröter & Sievers, 1971), and the aggregation and disaggregation of microtubules can be influenced by changes in hydrostatic pressure (Tilney & Gibbins, 1969; Kitching, 1970).

Stewart *et al.* (1973) measured diurnal variations of water in differentiating secondary vascular tissues of *Eucalyptus regnans*. Differentiating xylem tissue experienced the greatest water stress between 8:00 p.m. and midnight; during this period the xylem tissue withdrew water from the developing phloem. These relatively high changes in water content in the vascular tissues probably had some modifying effects on the cytodifferentiation processes (Stewart *et al.*, 1973).

Borger & Kozlowski (1972) exposed germinating seeds and one-day-old seedlings of *Fraxinus* to various concentrations of polyethylene glycol for thirty-five days. Seedlings grown in twenty per cent polyethylene glycol had smaller periderm and xylem increments than control seedlings, whereas germinating seeds exposed to ten or twenty per cent polyethylene glycol failed to develop either periderm or secondary xylem during the treatment. Wright & Northcote

(1973) attempted to initiate cytodifferentiation in a sycamore callus by culturing explants on media containing various concentrations of polyethylene glycol and sucrose, respectively. Polyethylene glycol concentrations up to sixteen per cent produced a compact callus containing some areas of high mitotic activity, but none of the explants cultured on polyethylene glycol-containing media exhibited cytodifferentiation. The results of cytodifferentiation studies involving the use of this compound (Doley & Leyton, 1970; Doley, 1970; Borger & Kozlowski, 1972; Wright & Northcote, 1973) should, however, be viewed with skepticism because of the numerous toxic contaminants present in commercial preparations (Michel, 1966). Injury and certain unusual responses have resulted from its use (Lagerwerff, Ogata & Eagle, 1961; Jackson, 1962; Macklon & Weatherley, 1965; Leshem, 1966; Greenway, Hiller & Flowers, 1968), and toxic effects were not restricted to dialyzable impurities (Ruf, Eckert & Gifford, 1963; Greenway *et al.,* 1968).

Certain effects of water potential on cytodifferentiation could be related to tissue friability. A friable culture is one in which the cells are rather loosely connected and readily disassociated. Sussex & Clutter (1967) found that agar-grown *Eucalyptus* callus was compact and contained tracheary elements, but that this same tissue grown as a suspension culture lacked tracheary elements. The reaggregation and compression of the disassociated cells in cellophane bags was incapable of restoring cytodifferentiation; therefore Sussex & Clutter (1967) proposed the necessity of plasmodesmatal connections between the cells for the initiation of cytodifferentiation. An alternative explanation is that the disassociated cells in the liquid were more ‘leaky’ and that, either by secretion or leaching, the cells were incapable of mustering a sufficient concentration of endogenous metabolites in order to initiate cytodifferentiation. In cell suspensions of wild carrot, Steward, Mapes & Mears (1958) observed that tracheary elements usually formed from some of the central cells after the aggregates had reached a particular size. They suggested that cytodifferentiation in these cell clusters was dependent upon a critical size, so that the internal cells, destined to form tracheary elements, would be removed from the leaching action of the medium. Wilbur & Riopel (1971*a*,*b*) employed suspension cultures of *Pelargonium* that consisted of either single cells or aggregates of no more than 115 cells. After the cells were reaggregated in nylon cones, cytodifferentiation occurred (Fig. 32). A quantitative relationship existed between the cell population and the numbers of tracheary elements formed (Tables 2 and 3). Although it is impossible to determine from this experiment to what extent pressure (Brown & Sax, 1962; Brown, 1964) and conditioning of the medium were factors in the initiation of cytodifferentiation, it would be interesting to isolate some cone cells destined to form tracheary elements and attempt their culture in the absence of external pressure. Another approach might be to examine the medium. Assuming that the medium in the apex of the cone (xylogenic zone) had been conditioned by the differentiating cells, the cytodifferentiation properties of such a medium might be revealed

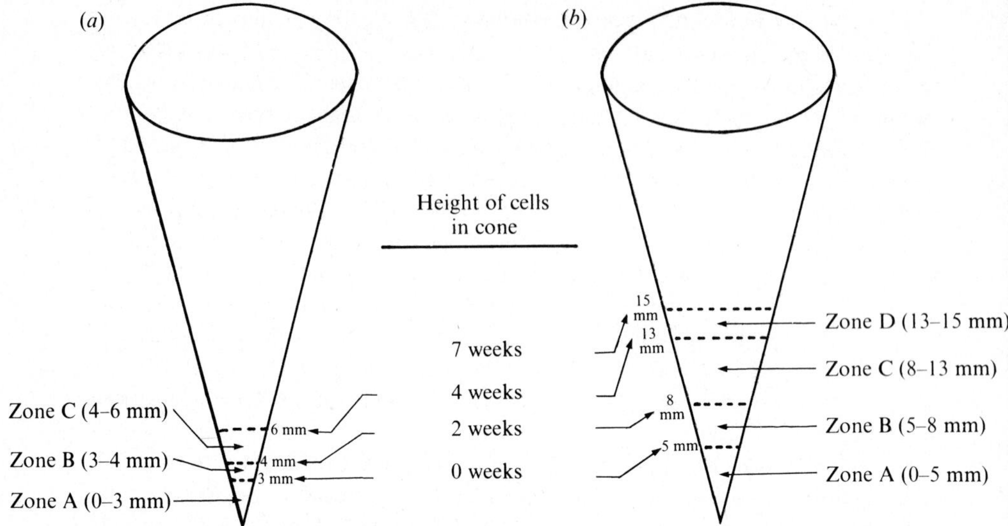

Fig. 32. Diagram of nylon cones employed for the cytodifferentiation of tracheary elements in reaggregated suspension cultures of *Pelargonium* cells. (*a*), 5000 cell cone; (*b*), 30 000 cell cone. The longitudinal and lateral distributions of tracheary elements were determined in the indicated zones after 2 (B), 4 (C) and 7 (D) weeks of growth. Zone A represents the region occupied by the initial inoculum. (From Wilbur & Riopel, *Bot. Gaz.* **132,** 1971*b*; courtesy of University of Chicago Press.)

by employing it as a culture medium with a freshly prepared suspension of *Pelargonium* cells that had not undergone the reaggregation treatment. Since tracheary elements always developed from small clusters of meristematic cells within the cone, plasmodesmatal connections between differentiating cells were probably present (Wilbur & Riopel, 1971*b*). Although there was no evidence that plasmodesmatal connections played any role in cytodifferentiation, the possible functional significance of these cytoplasmic interconnections is still not clear. Variations in the water potential may have been responsible for the differences in tracheary element elongation and alterations in secondary wall patterns observed by Beslow & Rier (1969) in cultured internodes of *Coleus,* since these experiments involved relatively high sugar concentrations.

Some recent reviews on water relations that may be of interest to the student of cytodifferentiation include Levitt (1972), Todd (1972), Livne & Vaadia (1972), Naylor (1972) and Hsiao (1973).

Temperature

Durzan *et al.* (1973) observed that xylogenesis occurred in cell aggregates in suspension cultures of white spruce (*Picea glauca* [Moench] Voss) under conditions of constant temperature (23.5 °C) and light. When the cultures were placed

TABLE 2. *Relationship between total number of reaggregated cells and longitudinal distribution of tracheary elements with an initial inoculum of 5000 cells* *

Total number of reaggregated cells/cone	Age of cone (weeks)	Height of cells in cone (mm)	Number of tracheary elements/1000 reaggregated cells		
			Zone A	Zone B	Zone C
5 000	0	3	0	—	—
10 000	2	4	0	—	—
38 000	4	6	9.7 ±2	4.9 ±1	0.7 ±1

* See Fig. 32(*a*). Within each zone the tracheary elements were determined in a radial region extending inward from the periphery of the cone for 700 μm. Estimations of the total number of reaggregated cells are taken from Wilbur & Riopel (1971*a*, fig. 8). Tracheary cell counts represent averages from data collected from over 50 sections taken from cells in cones from 2 to 4 experiments. Data include 95 % confidence limits. (From Wilbur & Riopel, 1971*b*; courtesy of University of Chicago Press.)

TABLE 3. *Relationship between total number of reaggregated cells and longitudinal distribution of tracheary elements with an initial inoculum of 30 000 cells* *

Total number of reaggregated cells/cone	Age of cone (weeks)	Height of cells in cone (mm)	Number of tracheary elements/1000 reaggregated cells			
			Zone A	Zone B	Zone C	Zone D
30 000	0	5	0	—	—	—
74 000	2	8	6.7 ±1	2.3 ±.6	—	—
190 000	4	13	37.2 ±6	14.4 ±4	10.2 ±2	—
391 000	7	15	39.7 ±8	31.8 ±6	26.9 ±5	10.7 ±2

* See Fig. 32(*b*). Within each zone the tracheary elements were determined in a radial region extending inward from the periphery of the cone for 700 μm. Estimations of the total number of reaggregated cells are taken from Wilbur & Riopel (1971*a*, fig. 8) with the exception of 391 000, which was calculated expressly for this experiment. Tracheary cell counts represent averages from data collected from over 50 sections taken from cells in cones from 2 to 4 experiments. Data include 95 % confidence limits. (From Wilbur & Riopel, 1971*b*; courtesy of University of Chicago Press.)

under conditions similar to those of a late-spring day, cytodifferentiation was not observed. In the latter case the cultures received cycles of fourteen hours light (23.5 °C) and ten hours dark (12 °C). The increased production of tannins in the cultures grown under the alternating conditions may have inhibited cytodifferentiation (see p. 30). However, some initiating event in xylogenesis, unrelated to tannin production, may have been inhibited by the low-temperature treatment. Explants of Jerusalem artichoke readily exhibited tracheary element cytodifferentiation, but 17 °C was the minimum temperature at which xylem differentiation occurred (Gautheret, 1961*b*). Denne (1971), on the other hand, observed tracheid production, with relatively thick secondary wall formation, in seedlings of *Pinus sylvestris* grown on a day/night temperature cycle of 17.5/10 °C. The minimal temperature for cytodifferentiation probably varies considerably among plants and is genetically controlled.

The incubation of certain tissue cultures at relatively high temperatures can profoundly affect both the extent and morphology of differentiating tracheary elements. Explants of Jerusalem artichoke tuber (*Helianthus tuberosus* L.) formed a genuinely compact wood at 31 °C (Gautheret, 1961*b*). Naik (1965) stated that in Jerusalem artichoke explants, cultured for 28 days at 35 °C, approximately 43 per cent of the total cell population formed tracheary elements! Why do temperatures of 30 °C and higher cause such a tremendous stimulation in the initiation of the process? The rate of some metabolic reaction critical to the initiation process may have a temperature optimum within this range, or perhaps there is an enhanced production of some xylogenic substance at the higher temperatures. A strain of tobacco callus was apparently capable of the biosynthesis of cytokinin at relatively high temperatures, but the pathways leading to the formation of cytokinins were blocked at lower temperatures (Syōno & Furuya, 1971). Unfortunately, no one has studied the underlying factors involved in this unusual temperature effect on cytodifferentiation.

Light

Relatively little research has been reported on the effects of light on the differentiation of tracheary elements. Some effects may be a reflection of the influences of light on cell division as a prerequisite for cytodifferentiation. White light inhibited cell division in explants of Jerusalem artichoke, but the response was only effective prior to DNA synthesis and in the presence of 2,4-D (Yeoman & Davidson, 1971). Mizuno, Komamine & Shimokoriyama (1971) examined the effect of light on xylogenesis in cultured slices of carrot root phloem. When explants from cultivars 'Nakamura-senkô-futo' and 'Yamada-hyakunichi-senkô-naga' were precultured in the dark for two days on a modified Murashige & Skoog (1962) medium and subsequently transferred to a medium containing 2,4-D (5×10^{-6} M) in the dark, no tracheary elements differentiated. When the precultured slices were transferred from the dark to a similar 2,4-D medium in the light, extensive xylogenesis occurred. Since the addition of cytokinins to the 2,4-D medium produced tracheary elements under both light and dark conditions, the authors postulated that a cytokinin-like substance, active in xylogenesis, was produced in the tissues of these cultivars by the action of the light. Explants from cultivars 'Kuroda-gosun', 'Kintoki' and 'Kokubu-senkô-onaga' produced extensive tracheary element formation following dark culture on a 2,4-D medium; the initial explants from these cultivars evidently contained sufficient endogenous cytokinins for the initiation of xylogenesis (Mizuno *et al.*, 1971). An increased number of tracheary elements were formed in the hypocotyl of

Sinapis alba L. in the presence of the far-red-absorbing form of the phytochrome pigment (P_{730}), although the pattern of cytodifferentiation within the bundles and the course of the bundles within the hypocotyl were identical in etiolated and illuminated seedlings (Kleiber & Mohr, 1967). So far this report on the involvement of phytochrome in tracheary element differentiation has remained unconfirmed. The presence of white light stimulated xylogenesis in isolated stem segments of *Coleus*, but there was no indication that flashes of either red or far-red light had any appreciable effect on xylogenesis in similar cultures incubated in the dark (E. Fosket, unpublished observations). If the disposition of microtubules can be influenced by a photoreceptor, microfibril orientation in the developing secondary wall could be altered by monochromatic light absorbed by the photoreceptor. Miller & Stephani (1971) suggested that various wavelengths of light may alter microtubule orientation and, in turn, affect cell wall growth. The orientation of microtubules in the apical cells of fern gametophytes was reported to be different after treatment with red light as compared to blue light (Stetler & DeMaggio, 1972). Some preliminary experiments have suggested that blue and far-red wave lengths may influence the pattern of secondary wall deposition in differentiating tracheary elements (Baba & Roberts, unpublished observations; Fig. 4). A critical and definitive study of the qualitative effects of light on cytodifferentiation is seriously needed to either refute or confirm the involvement of photoreceptors functioning in cytodifferentiation.

Mechanical stress

The application of mechanical stress, physical restraint, geotropic stimulation, and thigmotropism all have one feature in common: these are traumatic events that initiate the release of ethylene from the stimulated tissue (Abeles & Abeles, 1972). Consequently, any cytodifferentiation response arising from these stimuli probably will be influenced, to some extent, by the ethylene released within the tissue. Trauma-induced ethylene biosynthesis has been reported in secondary xylem tissue (Cooper, 1972). The application of mechanical stress to branches of white pine, apple and peach resulted in increases of ethylene in the internal atmosphere of the wood by over fifty per cent (Leopold, 1972). Leaf epinasty resulting from clinostat treatment was associated with increased ethylene production (Lyon, 1972; Leather *et al.*, 1972); enhanced xylogenesis in wounded *Coleus* shoots subjected to clinostat treatments (Roberts & Fosket, 1962) may have involved stress-induced ethylene biosynthesis. According to Kennedy (1970), differentiating secondary xylem, in common with other crystalline materials, displays a piezoelectric potential under stress. Differences in stem form associated with changing stress patterns, e.g., in reaction wood * formation, may be due to

* Reaction wood refers to a special type of secondary xylem produced on the lower sides of leaning stems of conifers (compression wood) and on the upper sides of geotropically stressed branches of dicotyledons (tension wood) (see Esau, 1965*a*).

an altered distribution of carbohydrates and hormones stimulated by potentials developed piezoelectrically. Although this interesting idea reappears periodically (see review by Schrank, 1959, for earlier literature), there is no solid evidence to support the view that the transport of plant hormones *in vivo* is in any way regulated by bioelectric potentials. The thigmotropic response of *Passiflora caerulea* involves extensive lignification of the xylem; this same response was imitated with exogenous auxin (Reinhold, Sachs & Vislovska, 1972). The authors failed to determine whether the thigmotropic response involved only the lignification of existing tracheary elements or whether the stimulus initiated cytodifferentiation in addition to the effect on lignification (T. Sachs, personal communication). The effects of mechanical stress on the orientation of cell division in callus cultures (Yeoman & Brown, 1971) will be of interest to future workers involved with the effects of mechanical stress on cytodifferentiation.

Electromagnetic effects

The treatment of cells initiating cytodifferentiation in root meristems of *Allium* and *Narcissus,* with magnetic fields (40–360 $\times$ 10^3 amp/sec) generated by permanent magnets obtained from magnetron tube assemblies, revealed some remarkable abnormalities in tracheary element differentiation (Dunlop & Schmidt, 1964, 1965, 1969). All protoxylem elements exhibited some form of scalariform-reticulate secondary wall pattern; frequently the secondary wall thickenings were oriented in broad longitudinal bands instead of the typical transverse patterns. Ultrastructural studies have not been made on tracheary element differentiation under these experimental conditions, and it would be of great interest to observe the orientation of microtubules in differentiating tracheary elements during electromagnetic field treatment.

Oxygen

The dissolved oxygen concentration was reported to regulate the pathway of cytodifferentiation in suspension cultures of carrot tissue. Below a critical level of oxygen, embryogenesis was initiated, whereas above this concentration rhizogenesis was favored (Kessell & Carr, 1972).

Carbon dioxide

Bradley & Dahmen (1971) observed certain effects of carbon dioxide, employed in combination with 2,4-D and kinetin, on cytodifferentiation in cultured peach mesocarp tissue. In comparison with control callus grown under normal atmospheric conditions, the increased levels of carbon dioxide were associated with fragile cell wall development, enhanced starch synthesis, and an extremely high incidence of cytodifferentiation in the form of phloem and xylem elements. The tracheary elements formed under these experimental conditions possessed a spiral pattern of secondary wall thickening instead of the typical scalariform-reticulate pattern (Bradley & Dahmen, 1971). Plant tissue cultures typically release ethylene (Gamborg & LaRue, 1971; LaRue & Gamborg, 1971), and carbon dioxide has an antagonistic effect on the physiological responses due to ethylene (Burg & Burg, 1967). Because extremely low concentrations of ethylene were stimulatory to cytodifferentiation (Roberts, unpublished observations), these observations suggest that the exogenous carbon dioxide was possibly effective in reducing the physiologically active concentration of ethylene to a range responsible for these effects (see Chapter 3). An alternative possibility is that carbon dioxide *per se* has some direct stimulatory effect on cytodifferentiation. The results reported by Bradley & Dahmen (1971) involving exogenous ethylene (5 ppm) cannot be considered valid because of the toxic effects of ethylene at this concentration (R. W. Zobel, personal communication). Similar experiments should be undertaken in which atmospheric ethylene, both from endogenous and exogenous sources, is carefully monitored in order to conduct a critical study of the relative effects of carbon dioxide and ethylene on cytodifferentiation. The effects of carbon dioxide on xylogenesis may in some manner be related to the effect of carbon dioxide in mimicking auxin-induced growth responses (Rayle & Cleland, 1970).

Ozone

The differentiation of tracheary elements in excised and wounded stem internodes of *Coleus,* cultured on a sucrose–IAA medium, was strongly inhibited in the presence of ozone (50 ppm), and the effect may be due to the oxidation of xylogenic auxin (Rier & Owens, 1973).

In summary, the following statements apply to the effects of external environmental influences on cytodifferentiation:

1. The most effective concentration of xylogenic hormones for cytodifferentiation *in vitro* may be dependent, to some extent, on the water potential of the nutrient medium. The effectiveness of hormonal levels on cytodifferentiation *in vivo* may vary with the water potential of the differentiating vascular tissues. Cellular water deficits may seriously affect organelles involved in various stages of cytodifferentiation.
2. Cytodifferentiation results obtained by varying the water potential of the external medium with polyethylene glycol should be viewed with caution, because of the presence of potentially toxic contaminants.
3. Cytodifferentiation in suspension cultures may be induced by 'packing' and reaggregating the cells in sterile nylon cones. The successful induction of

cytodifferentiation under these conditions may result from one or more of the following: (*a*) increased wall pressure by high cell density, (*b*) medium conditioning by leaky cells, (*c*) restricted gaseous exchange, and (*d*) plasmodesmatal connections.

4. Xylogenesis *in vitro* may be suppressed by cyclic environmental conditions involving low temperature treatments. Xylogenesis in some systems *in vitro* is highly stimulated by temperatures between 30 and 35 °C; the reason is unknown.

5. The effects of light on cytodifferentiation have not been examined critically. However, light may influence xylogenesis by one or more of the following photoregulatory mechanisms: (*a*) influencing cell division, (*b*) regulation of cytokinin biosynthesis, (*c*) phytochrome interaction, and (*d*) by some unknown photoreceptor involved in microtubule and microfibril orientation.

6. Some cytodifferentiation effects resulting from mechanical stress probably involve physiological responses arising from the release of trauma-induced ethylene production.

7. Cytodifferentiation in the presence of experimentally induced electromagnetic fields results in aberrant secondary wall patterns.

8. The stimulation of xylogenesis by carbon dioxide in cultured peach mesocarp tissue may involve antagonism with endogenous ethylene or may involve a direct response to atmospheric carbon dioxide.

9. Ozone was reported to inhibit xylogenesis. Oxygen has no known effect on cytodifferentiation but may influence organogenesis.

9

Chemical inhibitors and cytodifferentiation

Most research workers engaged in studies on cytodifferentiation have, at some time during the course of their experiments, resorted to the use of chemical inhibitors. A detailed discussion of the disadvantages of the use of chemical inhibitors in developmental studies is beyond the scope of this chapter, but a few words of caution are in order. The underlying basis for the physiological responses resulting from the use of some inhibitors is not well understood, and assumptions concerning the mode of action of a given inhibitor are frequently made that have little or no experimental justification. Rare indeed is the chemical inhibitor that is truly specific (and freely reversible!) and does not have a spectrum of side effects. It is doubtful, in fact, if the chemical disruption of any single metabolic pathway can be considered an 'isolated event' without unforeseen metabolic consequences.

Chemical inhibitors that have been employed in studies on tracheary element differentiation have generally fallen into three categories:

1. Inhibitors of some reaction or event considered as a possible prerequisite for a target cell to enter into a period of competence for the initiation of cytodifferentiation, e.g., DNA synthesis by FUdR and hydroxyurea, spindle formation in mitosis by colchicine, cytokinesis by caffeine, and DNA-dependent RNA transcription by actinomycin D.
2. The inhibition of some hormonal prerequisite for xylem differentiation, e.g., inhibition of GA biosynthesis by AMO-1618 and CCC, ethylene antagonism by carbon dioxide and TH-6241, and auxin transport blocking by TIBA, DPX-1840 and morphactins.
3. The selective inhibition of some event occurring after the initiation of cytodifferentiation, e.g., disruption of cellular microtubules by colchicine.

The blocking of a hormonal prerequisite for cytodifferentiation can involve blocking the biosynthesis of the hormone, antagonism at the presumed site of action, or the suppression of some physiological characteristic of the hormone. In the latter case, the inhibition of polar auxin transport profoundly affects cytodifferentiation. The possible relationship between the two processes is unknown.

For more detailed information concerning the applications of specific metabolic inhibitors, the reader is advised to consult the many comprehensive reviews, e.g., D'Amato (1960), Gelfant (1963), Hochster & Quastel (1963), Webb (1963, 1966), Kihlman (1966), Deysson (1968), Bücher & Sies (1969) and Roy-Burnman (1970).

Anti-ethylene compounds

Some attempts have been made in cytodifferentiation studies to employ carbon dioxide as a competitive antagonist of ethylene (Radin & Loomis, 1969; Bradley & Dahmen, 1971), since carbon dioxide is known to have an antagonistic effect on physiological responses resulting from ethylene treatment (Burg & Burg, 1967; see Chapter 8). Radin & Loomis (1969) reported that cambial activity in cultured radish roots was almost completely eliminated with 1.1 ppm ethylene, and that the inclusion of one per cent carbon dioxide had little antagonistic effect on the ethylene response. Higher concentrations of carbon dioxide were not tested. Bradley & Dahmen (1971) found a remarkable enhancement of vascular differentiation in peach mesocarp callus cultured on a 2,4-D plus kinetin medium with atmospheric carbon dioxide added. The extent of cytodifferentiation increased with increasing concentrations of carbon dioxide from one per cent to eleven per cent. In the experiment of Bradley & Dahmen (1971), however, exogenous ethylene was *not* added to the system, but the added carbon dioxide probably antagonized ethylene released by the callus. In employing carbon dioxide as an ethylene antagonist, it is virtually impossible to determine the most effective concentration ratio of the two gases in order to achieve optimum antagonistic responses. This is because the concentration of atmospheric ethylene within the sealed culture vessel constantly increases during the incubation period as the gas is released from the callus source. It is hoped that more precise techniques can be developed for employing carbon dioxide as an ethylene antagonist.

Some antagonism to ethylene-induced responses has been demonstrated with 5-methyl-7-chloro-4-ethoxycarbonylmethoxy-1,2,3-benzothiadiazole (TH-6241), but the mode of action of this compound has not been elucidated to date (Parups, 1973).

Benzyl isothiocyanate appears to play a role in the endogenous regulation of ethylene biosynthesis in papaya (*Carica papaya* L.) fruit (Patil & Tang, 1974). The enzymatic hydrolysis of benzyl glucosinolate in the pulp tissue of the fruit leads to the release of benzyl isothiocyanate. The latter compound may suppress the biosynthesis of the ethylene precursor methionine (Patil & Tang, 1974). Since benzyl isothiocyanate is a natural plant product, developmental studies involving the control of ethylene production should be undertaken with this compound.

Anticytokinins

The development of an anticytokinin would be a useful research technique, but the development of such an inhibitory agent must await further progress concerning the nature of endogenous cytokinins involved in cytodifferentiation; the origin, metabolism and site of action of these substances is poorly understood at this time (see review by Hall, 1973).

Antigibberellins

Gibberellin antagonists have been shown to alter xylem differentiation (Macchia, 1967). The application of AMO-1618 and CCC, respectively, to cultured *Pisum* seedlings resulted in decreased vessel size and lignification; apparently the mitotic activity of the shoot apical meristem was unaffected by the treatment (Macchia, 1967). Macchia (1967) concluded that the antigibberellin treatments had affected cell metabolism beyond the inhibition of gibberellin biosynthesis. Siebers & Ladage (1973) employed AMO-1618 and CCC in a study of the development of the interfascicular cambium in young castor bean (*Ricinus*) hypocotyls. The results obtained were inexplicable, since the inhibition of cambial development induced by either AMO-1618 or CCC gave no recovery on the addition of GA. On the contrary, the inhibitory effects of the growth retardants were intensified! An experiment should be conducted employing these inhibitors with cultured explants of Jerusalem artichoke tuber. This tissue system produced a marked synergism to auxin–cytokinin-induced cytodifferentiation by the addition of GA (Dalessandro, 1973*b*). We should be able to determine the effects of these inhibitors on the GA-induced xylogenic response with more precise regulation of the cytodifferentiation process than possible in the previous studies.

Inhibition of auxin by TIBA

Numerous studies have been conducted on the effects of 2,3,5-triiodobenzoic acid (TIBA) on cytodifferentiation (see review by Roberts, 1969). There is a close similarity between the physiological responses given by TIBA, morphactins and 1-naphthylphthalamic acid (Schneider, 1970). How TIBA acts is not precisely known, but it may involve blocking auxin transport through sulfhydryl inactivation (Niedergang-Kamien & Leopold, 1957), through IAA degradation by increased IAA-oxidase activity (Audus & Bakhush, 1961), or perhaps through auxin immobilization (Christie & Leopold, 1965; Keitt & Baker, 1966; Winter, 1967). Earlier studies had indicated that TIBA lowered the amount of diffusible auxin within the plant (Audus & Thresh, 1956; Winter, 1968; Cummings, 1959). In any case, the experimental results obtained with TIBA are consistent with the idea that there is less endogenous auxin available for cytodifferentiation processes.

Possibly, TIBA inactivates some endogenous sulfhydryl requirement for cytodifferentiation. Roberts & Sankhla (1973) reported that auxin–cytokinin-induced xylogenesis in explants of lettuce pith was completely blocked by the incorporation of TIBA in the medium, but that in the presence of TIBA plus cysteine a slight xylogenic response was observed.

Cambial development and cytodifferentiation in isolated hypocotyls of castor bean seedlings invariably starts at the basal end of the cultured explants and extends in an acropetal direction (Siebers & Ladage, 1973). Since this pattern of development could be a reflection of the movement of endogenous auxin, Siebers & Ladage (1973) examined the effects of TIBA on this phenomenon. In the presence of TIBA (10 ppm) cambial activity was doubled, and the differentiation of xylem elements was markedly enhanced, compared to untreated controls. However, at 50 ppm TIBA, cambial activity and cytodifferentiation were significantly inhibited. The formation of callus at the apical cut surface and the complete inhibition of root initiation at the basal cut surface supported the concept that TIBA had an inhibitory effect on the basipetal transport of endogenous auxin (Siebers & Ladage, 1973).

Robnett & Morey (1973) found that the application of TIBA to erect mesquite (*Prosopis*) seedlings produced decreased numbers of tension wood fibers concomitant with a significant increase in the amount of axial xylem parenchyma. Tension wood fibers were found to be a predominant cell type in the secondary xylem of *Prosopis*. The TIBA treatment produced parenchyma cells that were thick-walled, fusiform and filled with large amounts of starch. Cell division appeared unaffected and the cambial zone remained meristematic. The effect of TIBA on favoring the differentiation of parenchyma cells from cambial fusiform initials probably resulted from a reduction of endogenous auxin (Robnett & Morey, 1973). Shininger (1971) also found that fusiform initials formed parenchyma cells under conditions of low auxin. Krause (1971) examined histological changes in the cambium and shoot meristems of soybeans treated with TIBA. In the upper internodes of treated plants, in comparison with untreated controls, there was increased procambial activity and the formation of relatively small vessels. The reduction in size of vessels differentiated in the presence of TIBA was observed also by Ghorashy (1967).

There was a similarity between the observed effects of TIBA on cytodifferentiation from fusiform initials in *Prosopis* and the effects of the auxin transport inhibitor DPX-1840 on the same plant (Morey, 1973). The latter compound resulted in the differentiation of a relatively large fraction of the cambial derivatives to axial parenchyma cells instead of fibers. However, vessel member differentiation appeared unaffected. Similar to the results reported for TIBA-treated *Prosopis* seedlings (Robnett & Morey, 1973), these newly formed axial parenchyma cells were characterized by thick walls and a heavy deposition of starch. Morey (1973) suggested that the differentiation of axial parenchyma cells, instead of the typical xylem fibers in *Prosopis*, resulted from the relatively low auxin levels in the cytodifferentiation zone.

Morphactins

During the early 1960s, a group of chemically related compounds exhibiting growth-regulating properties was developed in the research laboratories of E.

Merck AG, Darmstadt, Germany. These compounds, derivatives of fluorene-9-carboxylic acid, were termed 'morphactins' because they were '*morph*ologically *act*ive substa*n*ce*s*' (Schneider, 1970). Nearly all of the research involving the effects of morphactins on cytodifferentiation has been conducted with chlorfluorenol (CFl; IT-3299; 2-chloro-9-hydroxyfluorene-9-carboxylic acid). For detailed information concerning the various physiological effects of these compounds the reader should examine the reviews by Schneider (1970), Ziegler (1970) and Ziegler & Sankhla (1974).

Pieniazek & Saniewski (1968) reported that CFl had a marked effect on the initiation of cambial activity and secondary xylem differentiation in excised dormant twigs of *Malus*. Although CFl alone had no apparent effect on the dormant cambium, in combination with benzyladenine there was a pronounced synergism. The newly formed secondary xylem consisted of short, wide tracheids instead of the long, narrow tracheids that are typically produced in *Malus*. In a continuation of this study, Pieniazek, Smolinski & Saniewski (1970) suggested that CFl exerted its morphological effects on secondary growth by interacting with endogenous auxin and cytokinins. N. Sankhla (personal communication) also observed the stimulation of cambial activity in *Petunia* seedlings treated with CFl; the resulting tracheary elements consisted entirely of small tracheids and xylem fibers. Cambial activity in the roots of *Pisum* and *Dolichos* was stimulated by CFl, particularly in the vicinity of the protoxylem poles; irregular vessels and short tracheids were produced (Saniewski, Smolinski & Pieniazek, 1968). Morphactin induced the formation of compression wood in *Picea excelsa* (Smolinski, Saniewski & Pieniazek, 1972). Kochhar, Anand & Nanda (1972) reported that CFl inhibited the differentiation of cambial derivatives into vessels in stem cuttings of *Salix tetrasperma*, but that cambial activity was apparently unaffected by the treatment. Since the possible effects of morphactins on cytodifferentiation in the absence of cambial activity has not been investigated, Roberts & Sankhla (1973) examined the induction of xylem differentiation in cultured explants of lettuce pith, cultured on a xylogenic medium containing IAA, kinetin and CFl. The incorporation of CFl (1 mg/1) in the medium strongly inhibited the xylogenic response in terms of numbers of tracheary elements formed, although the inhibitor had no discernible effect on the morphology of the differentiated cells. The inhibitory effect of CFl was partially reversed by the addition of cysteine (0.1 mM) to the CFl-containing medium; the suggestion was made that CFl blocked some sulfhydryl-containing system necessary for the initiation of xylogenesis (Roberts & Sankhla, 1973). The reversal of CFl inhibition by exogenous cysteine was confined entirely to tracheary element formation in the explant tissue adjacent to the nutrient medium. Xylogenesis in explants of lettuce pith cultured on an IAA–kinetin medium formed a ring of vertical strands of tracheary elements around the explant periphery (Dalessandro & Roberts, 1971); this expression of cytodifferentiation was completely inhibited by CFl and incapable of reversal by exogenous cysteine. Evidently the peripheral differentiation required transport for some distance, and this transport system remained blocked even with the addition of cysteine. Krelle (1970) indicated that morphactins may involve sulfhydryl inactivation, since he observed that CFl inhibited the activity of the sulfhydryl-containing enzyme urease, and that the application of exogenous cysteine reversed the inhibition. In addition, Krelle (1970) found that the effect of CFl on adventitious root formation in cuttings of *Phaseolus* hypocotyl was reversed by exogenous cysteine. The data of Roberts & Sankhla (1973) supported the view that xylogenesis in lettuce pith required some sulfhydryl-containing system, not associated with auxin transport, and that this system was reversibly blocked by CFI. The relationship of cell division to xylogenesis has been discussed at length (see Chapter 4). Morphactins are inhibitors of cell division activity (Ziegler, 1970) and sulfhydryl groups are functional in the mitotic process (Jocelyn, 1972). Microtubules of the mitotic apparatus contain disulfide linkages (Mazia, 1967) and these organelles may be inactivated by CFl. The observation that CFl reoriented the spindle apparatus in dividing cells may be related (Ziegler, 1970). Morphactins inhibited the fusion of Golgi-derived vesicles during cell plate formation in cytokinesis (Ziegler, 1970), and this finding may be relevant to cytodifferentiation.

10

Epilogue

Nature has provided us with a remarkable array of unique cell types. Developmental studies similar to those of xylogenesis research may provide clues to the critical and controlling variables involved in the cytodifferentiation of these specialized cells.

One cell type that deserves further examination is the so-called 'phloeotracheid' (Benson, 1910). This cell presumably contains some of the structural and functional characteristics of both a sieve tube element and a tracheary element, and it was first observed by Benson (1910) in the haustorium of *Exocarpos* (Santalaceae). A similar type of cell was found in *Pedicularis* (Maybrook, 1917). Granule-containing vascular cells, presumed to be phloeotracheids, were observed in the haustorium of *Mida salicifolia* (Simpson & Fineran, 1970). These parasitic organisms may have evolved some unknown hormonal stimulus that is capable of the induction of a unique cell type found only in these host–parasite relationships.

Developmental studies involving plants exhibiting anomalous cambial activity may yield information on the metabolic requirements for the differentiation of secondary xylem (see Philipson *et al.*, 1971). In the stem of *Doxantha unguis-cati* (Bignoniaceae), four arcs of the normally functioning vascular cambium either cease or reduce the cytodifferentiation of xylem elements, while continuing the production of secondary phloem (Dobbins, 1971; Figs 33, 34). Each of these cambial arcs, associated with unidirectional activity, is bordered by multiseriate rays. These rays appear to be involved, in some manner, in the induction of the unidirectional cambial activity (Dobbins, 1971). In the stem of *Pyrostegia venusta* a similar condition exists; four sections of the vascular cambium form considerable phloem and relatively little xylem (Philipson *et al.*, 1971). Another variation from normal cambial activity involves the production of alternating bands of different cell types within the secondary xylem tissue. In *Hibiscus syriacus* the secondary xylem consists of bands of lignified vessels and fibers alternating with nonlignified parenchymatous tissue (Belan, 1970). A single growth ring of *Hoheria* may consist of twenty or more bands of fibers alternating with layers of vessels and parenchyma (Philipson *et al.*, 1971). What is the hormonal signal that switches cytodifferentiation from one cell type to another within the same tissue, and how is this signal regulated?

A different approach that needs investigation consists of making a comparison

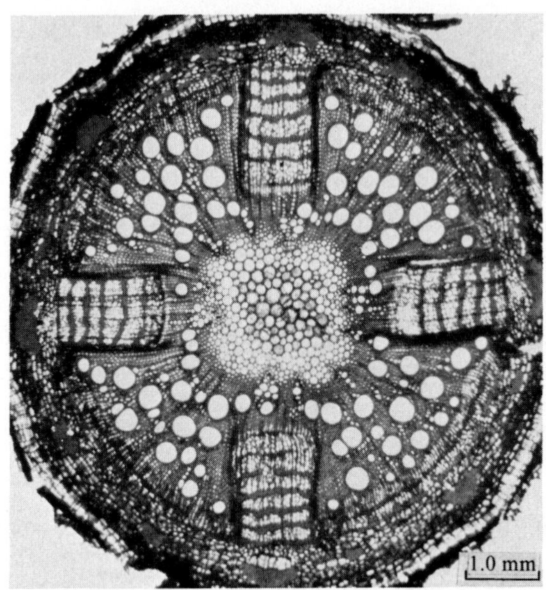

Fig. 33. Transverse section of stem of *Doxantha unguis-cati* showing anomalous cambial activity. Four arcs of the vascular cambium form little or no secondary xylem but considerable secondary phloem. As secondary growth continues, the bidirectional and unidirectional arcs of the cambium become separated, and radial splits occur between the furrows of secondary phloem (unidirectional activity) and the normal production of secondary xylem (bidirectional activity). (Courtesy of D. Dobbins.)

of the CDS of similar cell types, e.g., foliar sclereids of *Camellia japonica* and wound vessel members. Both of these cell types differentiated following a wound reaction, and both types successively demonstrated nuclear enlargement, cell expansion and extensive secondary wall synthesis (Foard, 1960; see Foard, 1970). The formation of foliar sclereids in *Camellia* was induced following wounding only in immature expanding leaves; this suggests that IAA may be a hormonal requirement for sclereid cytodifferentiation. Although explants excised from immature leaves failed to form sclereids following the prolonged culture on a sucrose-containing medium (Foard, 1960), experiments on elucidating the hormonal requirements for the induction of sclereid cytodifferentiation were not done. The possibility of the experimental induction of both sclereids and wound vessel members within the same tissue system is an intriguing idea.

Another cell type that deserves attention is the xylem transfer cell (Wooding & Northcote, 1965; Wooding, 1969; Pate, Gunning & Milliken, 1970). Recently a detailed examination was made concerning the ontogeny of xylem transfer cells in *Hieracium floridundum* (Yeung & Peterson, 1974). The CDS of these cells varied with the type of tracheary element associated with the differentiating transfer cell. Parenchyma cells associated with protoxylem and metaxylem ele-

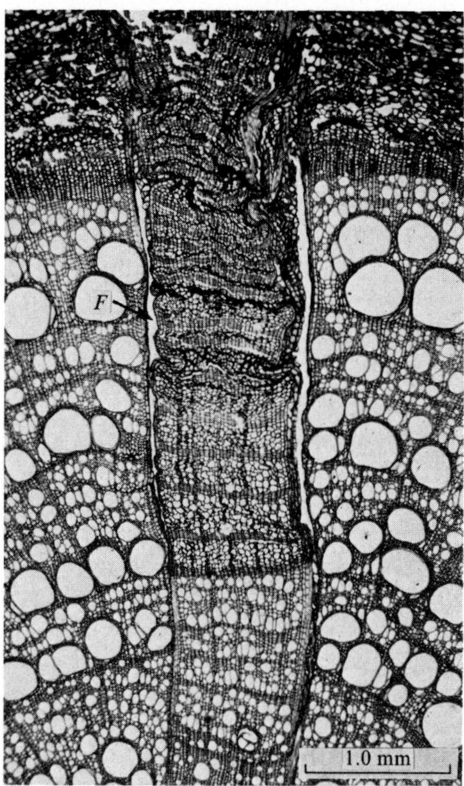

Fig. 34. Transverse section of stem of *Doxantha unguis-cati* showing the effect of unidirectional cambial activity. The secondary phloem becomes crushed, and the 'furrow phloem' has sheared along the rays, resulting in fissures (*F*). (From Dobbins, *Am. J. Bot.* **58,** 1971.)

ments, in the vicinity of the rhizome apex, usually divided prior to the formation of secondary wall ingrowths. Following division, the daughter cell next to the tracheary element differentiated as a transfer cell, and the other daughter cell formed a xylem parenchyma cell. Transfer cells associated with protoxylem cells formed the most extensive wall ingrowths, whereas transfer cells associated with late metaxylem elements had little or no wall ingrowths. Parenchyma cells contiguous with secondary wall elements completely lacked wall ingrowths (Yeung & Peterson, 1974). This ontogenetic study suggested that the differentiating transfer cells received substances from the xylem necessary for wall ingrowth formation; the more efficient the xylem as a conducting system, the less capable the tissue was in furnishing hormones or metabolites involved in wall synthesis. Wall ingrowth formation, however, was not entirely mediated by external influences, since the extent of wall formation was closely related to the preceding mitotic activity (Yeung & Peterson, 1974). The cytodifferentiation of transfer cells may be a hormonal response similar to xylem differentiation, and perhaps transfer cells represent 'aborted tracheary elements' (Wooding & Northcote, 1965). Following the onset of secondary growth in the rhizome of *Hieracium*, the wall ingrowths disappeared, and the mature transfer cells remained unlignified for a considerable length of time (Yeung & Peterson, 1974). The latter events clearly are not typical of the ontogeny of tracheary elements.

The cytodifferentiation of sieve tube elements presents some perplexing problems. The study of this cell type requires special techniques for the cytochemical demonstration of callose deposits. Few developmental botanists have made comparative studies of the cytodifferentiation of xylem and phloem under the same experimental conditions. Whether or not both cell types are invariably present in cultures maintained under xylogenic conditions is a point that has not been seriously investigated. Sieve tube element regeneration in *Coleus*, similar to wound vessel member formation in this system, was apparently limited by IAA (LaMotte & Jacobs, 1963; see review by Jacobs, 1970*a*). Gibberellins (DeMaggio, 1966) and cytokinins (LaMotte & Houck, 1973) may play roles in the cytodifferentiation of sieve tube elements. The interrelationships between various sugar : auxin ratios and the resulting vascular arrangements and cell types in callus cultures has never been adequately explained (Wetmore & Rier, 1963). The time course of phloem and xylem regeneration around a *Coleus* stem wound revealed that the first developing sieve tube elements appeared two days after wounding, and approximately twenty-four hours in advance of the first observable wound vessel members (Thompson, 1967). This 'phloem–xylem sequence' has been observed, under natural conditions, in a large number of different plant groups (Esau, 1965*b;* see Chapter 5). So far, the possibility of a causal relationship between the CDS of the sieve tube element and the initiation of the CDS in the xylem element has not been investigated. Although distinct differences in the metabolism of the two cell types are apparent, lignified sieve elements have been observed in the wheat leaf (Kuo & O'Brien, 1974). The requirements for phloem and xylem differentiation are evidently different in the lower plants. Both the trumpet hyphae of the Laminariales and the sieve tubes of *Macrocystis* and *Nereocystis* contain callose deposits and are functional in conduction, yet tracheary elements have never been observed in any of the brown seaweeds (Fritsch, 1945). The lack of auxin responses in certain algal groups may involve differences in the cytological arrangement of microtubules and in biochemical differences in cell wall metabolism (Jacobs, 1970*b*).

It is puzzling that, with the exception of xylem and phloem, other types of specialized plant cells have seldom been induced to form in plant tissue cultures. Would it be possible, with the proper cultural conditions, to induce populations of root hair cells, stomatal guard cells, collenchyma or sclereids in our cell cultures (Torrey, 1971)? Is the cytodifferentiation of each of these cell types initiated by a special hormonal combination? Our ability to regulate experimentally the initiation of a given CDS will be considerably enhanced with the development of synchronous populations of cell-suspension cultures. One of the most exciting prospects concerns the utilization of the techniques developed in protoplast isolation and in the regeneration of cell cultures from protoplasts (Maretzki & Nickell, 1973). As we discussed earlier (see Chapter 4), cytodifferentiation was readily induced in cortical explants of pea root following endoreduplication (Phillips & Torrey, 1973). Cortical explants of pea root have also been employed for the preparation of isolated protoplasts in microculture (Landgren & Torrey, 1973). These protoplasts regenerated a cell wall, and metaphase figures indicated the presence of diploid, tetraploid and octoploid divisions. If these isolated meristematic cells could be induced, by suitable manipulations of the medium, to initiate cytodifferentiation, one of our long-sought goals in developmental physiology would be realized (Roberts, 1969). The entire CDS could then be subjected to continuous microscopic examination and cytochemical analysis.

Some recent developments

In a re-examination of cambial terminology, Butterfield has suggested that the term 'cambium' be defined as 'a multiserate zone of periclinally dividing cells lying between the differentiating secondary xylem and phloem, with a distinct initial capable of both periclinal and anticlinal divisions lying somewhere within each radial file of cells' (1975, *IAWA Bull.* pp. 13–14). The hormonal regulation of cambial activity has received further attention from developmental physiologists. Zajaczkowski & Wodzicki examined the production of secondary xylem in sterile cultures of isolated stem segments (*Pinus silvestris*) fed with synthetic media and it was found that cytodifferentiation continues only for approximately four to six weeks (1975, *Physiol. Plant.* **33,** 71–4). Enrichment of the medium with a methanolic extract from the cambial region of pine stem prolonged xylem production up to fifteen weeks, and the authors suggested that indoleacetic acid may act synergistically with some endogenous stimulant of cambial activity. In another study, a high rate of cambial division and radial expansion of the derivatives (*Pinus densiflora*) was not accompanied by high amounts of endogenous IAA, although a sharp decrease in IAA was observed when the radial diameter of the tracheids began to decrease (1974, *J. Japan Wood Res. Soc.* **20,** 507–11). Seasonal cambial activity in the latter study evidently was not associated with fluctuations of endogenous growth inhibitors, but abscisic acid has been detected in the bark tissues of *Taxus baccata* (1974, *C.R. Acad. Sci.* (Paris) **279,** 755–6). Jenkins & Shepherd have identified abscisic acid in stem extracts of *Pinus radiata* (1972, *New Phytol.* **71,** 499–509). Recently Jenkins showed that the injection of abscisic acid into stems of *Pinus radiata* seedlings (earlywood stage) resulted in a marked reduction in tracheid radial diameter and decreased mitotic activity of the cambium (1974, *Mechanisms of Regulation of Plant Growth,* pp. 743–50, Wellington, NZ). The implication is that endogenous abscisic acid levels, enhanced by summer moisture stress, may be an important factor in latewood anatomy and false ring formation. Determinations of IAA oxidase activity in extracts obtained from the cambial zone of *Pinus densiflora* indicated that enzymatic activity was higher in the latewood stage than in earlywood development (1974, *J. Japan Wood Res. Soc.* **20,** 512–15). Weidlich found that cambial activity in a petiole can be 'forced' to produce excessive amounts of secondary xylem and phloem by bridge grafting a petiole segment of Norway maple (*Acer plantanoides*) across the girdled shoot of a maple sapling (1974, *Can. J. Bot.* **52,** 1983–4). Parish examined, by freeze etching, the differentiating vascular cells in the cambial zone of crack willow (*Salix fragilis*) for a possible correlation between seasonal cambial activity and the sizes and frequencies of particles observed on the plasmalemma and tonoplast fracture faces (1974, *Cytobiologie* **9,** 131–43). The increased number of particles associated with the plasmalemma in the spring suggested that the particles may have had a physiological role in cell wall synthesis. In the Monocotyledoneae eccentric growth in horizontal stems of *Cordyline* and other woody Agavaceae revealed a close correspondence on the lower side of the stem between enhanced cambial activity and high levels of applied auxin, according to Fisher and coworkers. Experiments with [^{3}H]IAA applied to intact and excised horizontal stems of *Cordyline* showed that up to seven times more auxin accumulated in the cambial zone on the lower side compared to the upper side (1974, *Physiol. Plant.* **31,** 284–7; 1975, *Am. J. Bot.* **62,** 292–302). Reaction wood tracheids were not observed in the treated *Cordyline* stems, although the presence of reaction fibers in a monocotyledon (*Xanthorrhoea*) was recently reported by Staff for the first time (1974, *Protoplasma* **82,** 61–75). For additional information on the anatomy and physiology of reaction wood the reader should consult the comprehensive reviews by Scurfield (1973, *Science* **179,** 647–55) and Timell (see Bibliography and author index).

Studies on environmental and nutritional influences on cytodifferentiation have provided us with some interesting data. Light, temperature, and gibberellic acid, respectively, had inductive effects on the cytodifferentiation of vascular nodules and tracheary elements in cultured explants of Jerusalem artichoke (1974, *C.R. Acad. Sci.* (Paris) **279,** 1871–6). In the latter study, Saussay & Gautheret postulated that gibberellin biosynthesis in the cultures was regulated by light and temperature. Baas & Van der Graaff indicated that temperature was the environmental factor responsible for the modifications observed in the wood anatomy of numerous genera of vascular plants in relation to latitudinal and altitudinal distribution (1974, *IAWA Bull.* pp. 3–5). A 'hormone-like' effect of sucrose was illustrated in the further development of embryoids derived from cell suspensions of carrot (*Daucus carota*) root (1974, *Z. Pflanzenphysiol.* **74,** 85–90). Embryoids cultured in nutrient solutions containing one per cent sucrose produced normal plantlets, whereas embryoids grown in the presence of two per-cent sucrose produced only root growth and development. Bud morphogenesis has been induced by gamma irradiation in tobacco (*Nicotiana*) cultures (1973, *Radiat. Bot.* **13,** 381–3), although a similar treatment of orange (*Citrus sinensis*) ovular callus with gamma rays stimulated embryoid formation (1973, *Radiat. Bot.* **13,** 97–103).

Information on the possible relationship between the cell cycle and the initiation of plant cell differentiation is scarce, and the relevance of experimental results from animal cell cycles to similar phenomena in plant cells has not been established. According to Vonderhaar & Topper (1974, *J. Cell Biol.* **63,** 707–12) the terminal differentiation of mammary epithelial cells required some critical events which can occur only during a limited portion of the G_1 phase of the cell cycle. Consequently, certain cells previously arrested in the cell cycle must undergo proliferation as a prerequisite for arriving at this portion of the G_1 phase. The authors suggested that this cytodifferentiation requirement may be characteristic of other animal cell types. This finding is suggestive of the hypothesis made previously (see p. 51) that the initiation of the CDS in plant cells may involve the production of a specific RNA during some limited part of a cell cycle phase. In order for the cell to initiate cytodifferentiation the suggestion was made that it must recycle, perhaps repeatedly, for the necessary biosynthesis to occur within the transient period of the critical phase.

Since a shift in the metabolism from the biosynthesis of primary wall material to secondary wall development marks one of the stages in the cytodifferentiation of tracheary elements, considerable interest has been focused on this area. According to Noel (1974, *Ann. Bot.* **38,** 495–504), the helical pattern of secondary wall formation in tracheary elements appears analogous to the helical wall development in the velamen cells of the aerial root of the orchid *Ansellia gigantea*. Microtubules were observed to be associated with the developing helical strands. The development of the velamen cells resembles that of tracheary elements except that the former cells elongate radially, not longitudinally, and the profiles of the helical strands are asymmetrical and unpaired on adjacent walls. In another study the spatial relationship between microtubules and cellulose microfibrils was observed by Fujita and co-workers in differentiating tension wood fibers of *Populus euramericana* (1974, *J. Japan Wood Res. Soc.* **20,** 147–56). The orientation of microtubules coincided with the arrangement of cellulose microfibrils, and a shift in microtubule orientation was followed by a similar rearrangement of microfibrils during the developmental transition from the S_2 layer to the gelatinous layer. Heath has offered the hypothesis that cellulose synthetase complexes may be capable of movement in the plane of the plasmalemma. Possibly some unknown component serves to link and generate a sliding force between enzyme complex and microtubule, with the microtubule functioning as a guiding track for the system. Cellulose microfibrils spun out in the wake of the moving complexes would thus mirror the orientation of the microtubules (1974, *J. Theor. Biol.* **48,** 445–9). During tracheary element cytodifferentiation in lettuce (*Lactuca*) pith explants, Wright & Bowles detected alterations in the polysaccharide composition of the developing cell walls (1974, *J. Cell Sci.*, **16,** 433–43). Cytodifferentiation induced in the presence of zeatin was accompanied by nearly a twelvefold increase in the amount of labeled polysaccharide in isolated membrane fractions, and this change was apparent in the Golgi fraction at an earlier time than in the endoplasmic reticulum fraction. Biochemical experiments in progress may demonstrate a relationship between secondary wall formation during tracheary element cytodifferentiation and the enzymatic activity presumed responsible for the accumulation of the secondary wall monomers (G. Dalessandro, personal communication). Does the switching off of the activity of UDP-D-glucose 4-epimerase, UDP-D-glucuronic acid 4-epimerase, and UDP-L-arabinose 4-epimerase act as a signal for the differentiating cell to cease cell enlargement (primary wall synthesis) and initiate the deposition of a secondary wall? Progress to date on this problem has been hampered because of the difficulty in locating synchronous cell cultures suitable for both biochemical and xylogenic studies.

Recent experiments by Haddon & Northcote on the induction of vascular nodule formation in callus cultures initiated from bean (*Phaseolus vulgaris*) hypocotyl, illustrate some of the problems involved in cytodifferentiation studies (1975, *J. Cell Sci.* **17,** 11–26). The extent of cytodifferentiation was quantitatively expressed for xylem by lignin determinations with PAL activity and for phloem by measurements of callose formation with β-1$\rightarrow$3 glucan synthetase activity. PAL activity was maximum when the rate of vascular nodule formation was the highest, but the activity of callose synthetase was slightly different and subject to several interpretations. Different strains of bean callus, after several transfers, have shown considerable variation in retaining the capability of initiating cytodifferentiation. The authors suggested that bean callus may synthesize endogenous cytokinins, since the cultures did not require exogenous cytokinin for cell division and cytodifferentiation. An alternative explanation is that residual 2,4-D in the callus, carried over from the maintenance medium to the induction medium, may be substituting for the cytokinin requirement.

Some striking similarities exist between the cytodifferentiation of tracheary elements and the formation of unusual sclereid-like cells in the leaves of certain vascular plants. Lersten & Carvey have described the presence of dense masses of pitted and lignified cells in the distal periphery of ocotillo (*Fouquieria splendens*) leaves, and these unique cells apparently were intermediate in development between sclereids and tracheids (1974, *Can. J. Bot.* **52,** 2017–21). These lignified cells were formed primarily from enlarged bundle sheath cells and, to a lesser extent, mesophyll cells. Swany & Sivaramakrishna induced the formation of lignified cells, presumably similar to those observed in ocotillo leaves, by wounding stem internodes of selected monocotyledons (1972, *Phytomorph.* **22,** 305–14). The cytodifferentiation response, observed in *Commelina benghalensis* and *Gloriosa superba,* was observed within five days after wounding. The wound sclereids resulted either from a direct transformation of bundle sheath cells or from daughter cells following the division of bundle sheath cells. In sporadic cases, several layers of ground parenchyma adjacent to the wound site differentiated to form wound sclereids. The cells exhibited the helical, scalariform and reticulate wall thickenings associated with wound vessel members and experimentally induced tracheary elements.

Finally, the previous reports concerning the presence of cells described as 'phloeotracheids' have been confirmed and extended by Fineran (1974, *Ann. Bot.* **38,** 937–46). These cells were observed by scanning electron microscopy in the haustoria of several Santalaceous root parasites (*Exocarpus bidwillii, Mida salicifolia, Buckleya distichophylla, Comandra umbellata, Nestronia umbellata* and *Pyrularia pubera*). So far, physiological experiments on the cytodifferentiation and possible functional significance of these unusual cells have not been undertaken.

Bibliography and author index *

Abeles, A. L. & Abeles, F. B. (1972). Biochemical pathway of stress-induced ethylene. *Plant Physiol.* **50,** 496–598. *32, 101*

Abeles, F. G. (1972). Biosynthesis and mechanism of action of ethylene. *Annu. Rev. Plant Physiol.* **23,** 259–92. *31, 34*

Abeles, F. B. (1973). *Ethylene in plant biology*. New York: Academic Press. *35*

Abercrombie, M. (1967). General review of the nature of differentiation. In: *Cell differentiation,* ed. A. V. S. De Reuk & J. Knight, pp. 3–12. London: Churchill Ltd. *12, 42*

Alfieri, F. J. & Evert, R. F. (1965). Seasonal phloem development in *Pinus strobus. Am. J. Bot.* **52,** 626–7. *62*

Apelbaum, A., Fisher, J. B. & Burg, S. P. (1972). Effect of ethylene on cellular differentiation in etiolated pea seedlings. *Am. J. Bot.* **59,** 697–705. *32*

Audus, L. J. & Bakhush, J. K. (1961). On the adaptation of pea roots to auxins and auxin homologues. In: *Plant growth regulation,* ed. R. M. Klein, pp. 109–26. Ames: Iowa State University Press. *107*

Audus, L. J. & Thresh, R. (1956). The effects of synthetic growth-regulator treatments on the levels of free endogenous growth-substances in plants. *Ann. Bot. N.S.* **20,** 439–59. *107*

Avanzi, S., Cionini, P. G. & D'Amato, F. (1970). Cytochemical and autoradiographic analyses on the embryo suspensor cells of *Phaseolus coccineus. Caryologia* **23,** 605–38. *51*

Avanzi, S. & D'Amato, F. (1970). Cytochemical and autoradiographic analyses on root primordia and root apices of *Marsilea strigosa. Caryologia* **23,** 335–45. *51*

Avanzi, S., Maggini, F. & Innocenti, A. M. (1973). Amplification of ribosomal cistrons during the maturation of metaxylem in the root of *Allium cepa. Protoplasma* **76,** 197–210. *51*

Avery, G. S., Jr., Burkholder, P. R. & Creighton, H. B. (1937). Production and distribution of growth hormone in shoots of *Aesculus* and *Malus* and its probable role in stimulating cambial activity. *Am. J. Bot.* **24,** 51–8. *8, 59*

Bal, A. K. & Payne, J. F. (1972). Endoplasmic reticulum activity and cell wall breakdown in quiescent root meristems of *Allium cepa* L. *Z. Pflanzenphysiol.* **66,** 265–72. *77*

Balatinecz, J. J. & Kennedy, R. W. (1968). Mechanism of earlywood–latewood differentiation in *Larix decidua. Tappi* **51,** 414–22. *59, 63*

Ball, E. (1953). Hydrolysis of sucrose by autoclaving media, a neglected aspect in the technique of culture of plant tissues. *Bull. Torrey Bot. Club* **80,** 409–11. *90*

Ball, E. (1955). Studies on the nutrition of the callus culture of *Sequoia sempervirens. Ann. Biol.* **31,** 281–305. *7, 91*

* The numbers shown in italics at the end of each entry refer to the page at which this reference appears.

Banerjee, S. N. (1968). DNA synthesis in the root meristem cells of *Zea mays* dwarf mutants (d_1 and d_5). *Plant Cell Physiol.* **9,** 557–81. *47*

Banko, T. (1974). Some factors affecting cytokinin levels in the roots of *Coleus* and *Zea mays* L. PhD thesis, University of Idaho, Moscow, USA. *20*

Barlow, P. W. (1969*a*). Differences in response to colchicine by differentiating xylem cells in roots of *Pisum*. *Protoplasma* **68,** 79–83. *72*

Barlow, P. W. (1969*b*). Cell growth in the absence of division in a root meristem. *Planta* **88,** 216–23. *46*

Barlow, P. W. (1971). Properties of cells in the root apex. *Rev. Fac. Agron. La Plata* (Argentina) **47,** 275–301. *46*

Barlow, P. W. (1973). Mitotic cycles in root meristems. In: *The cell cycle in development and differentiation,* ed. M. Balls & F. S. Billett, pp. 133–65. London: Cambridge University Press.

Barnett, J. R. (1973). Seasonal variation in the ultrastructure of the cambium in New Zealand grown *Pinus radiata* D. Don. *Ann. Bot.* **37,** 1005–11. *78*

Basile, D. V., Wood, H. N. & Braun, A. C. (1973). Programming of cells for death under defined experimental conditions: relevance to the tumor problem. *Proc. Nat. Acad. Sci.* (USA) **70,** 3055–9. *8, 35, 45*

Beal, J. M. (1951). Histological responses to growth-regulating substances. In: *Plant growth substances,* ed. F. Skoog, pp. 155–66. Madison: University of Wisconsin Press. *7*

Belan, A. (1970). Differenciation rythmée des formations libéro-ligneuses secondaires, sous des conditions définies, chez l'*Hibiscus syriacus* L. et le *Tilia cordata* Mill. *C.R. Acad. Sci.* (Paris) **270,** 938–41. *111*

Benson, M. (1910). Root parasitism in *Exocarpos* (with comparative notes on the haustorium of *Thesium*). *Ann. Bot.* **24,** 667–77. *111*

Bergmann, L. (1964). Der Einfluss von Kinetin auf die Ligninbildung und Differenzierung in Gewebekulturen von *Nicotiana tabacum*. *Planta* **62,** 221–54. *7, 20*

Berlyn, G. P. (1970). Ultrastructural and molecular concepts of cell-wall formation. *Wood and Fiber* **2,** 196–227. *81, 82*

Berlyn, M. B., Ahmen, S. I. & Giles, N. H. (1970). Organization of polyaromatic biosynthesis enzymes in a variety of photosynthetic organisms. *J. Bact.* **104,** 768–74. *84*

Berlyn, M. B. & Giles, N. H. (1969). Organization of enzymes in the polyaromatic synthetic pathway: separability in bacteria. *J. Bact.* **99,** 222–30. *84*

Beslow, D. T. & Rier, J. P. (1969). Sucrose concentration and xylem regeneration in *Coleus* internodes *in vitro*. *Plant Cell Physiol.* **10,** 69–77. *7, 10, 98*

Bex, J. H. M. (1972). Effects of abscisic acid on nucleic acid metabolism in maize coleoptiles. *Planta* **103,** 1–10. *36*

Bierhorst, D. W. & Zamora, P. M. (1965). Primary xylem elements and element associations of angiosperms. *Am. J. Bot.* **52,** 657–710. *77*

Blum, J. L. (1941). Responses of sunflower stems to growth-promoting substances. *Bot. Gaz.* **102,** 737–48. *7*

Bonnemain, J.-L. (1971). Transport et distribution des traceurs après application de IAA-2-^{14}C sur les feuilles de *Vicia faba*. *C.R. Acad. Sci.* (Paris) **273,** 1699–702. *61*

Boothby, D. (1972). Uridine diphosphate-sugar metabolism by sycamore (*Acer pseudoplatanus*) cambial tissue. *Planta* **103,** 310–18. *93*

Borchert, R. (1972). Isoperoxidases as markers of the wound-induced differentiation pattern in potato tuber. *Develop. Biol.* **36,** 391–9. *86*

Borger, G. A. & Kozlowski, T. T. (1972). Effects of photoperiod on early periderm and xylem development in *Fraxinus pennsylvanica, Robinia pseudoacacia* and *Ailanthus altisima* seedlings. *New Phytol.* **71,** 691–702. *96, 97*

Borisy, G. G. & Taylor, E. W. (1967*a*). The mechanism of action of colchicine. Binding of colchicine-H3 to cellular protein. *J. Cell Biol.* **34,** 525–34. *71*

Borisy, G. G. & Taylor, E. W. (1967*b*). The mechanism of action of colchicine. Colchicine binding to sea urchin eggs and the mitotic apparatus. *J. Cell Biol.* **34,** 535–48. *71*

Bradley, M. V. & Crane, J. C. (1957). Gibberellin-stimulated cambial activity in stems of apricot spur shoots. *Science* **126,** 972–3. *8*

Bradley, M. V. & Dahmen, W. J. (1971). Cytohistological effects of ethylene, 2,4-D, kinetin and carbon dioxide on peach mesocarp callus cultured *in vitro. Phytomorphology* **21,** 154–64. *10, 35, 102, 103, 106*

Bragt, J. van & Pierik, R. L. M. (1971). The effect of autoclaving on the gibberellin activity of aqueous solutions containing gibberellin A_3. In: *Effects of sterilization on components in nutrient media,* ed. J. van Bragt, D. A. A. Mossel, R. L. M. Pierik & H. Veldstra, pp. 133–7. Wageningen: H. Veenman & Zonen N. V. *28*

Broughton, W. J. & McComb, A. J. (1971). Changes in the pattern of enzyme development in gibberellin-treated pea internodes. *Ann. Bot.* **35,** 213–28. *30*

Brown, C. L. (1964). The influence of external pressure on the differentiation of cells and tissues cultured *in vitro.* In: *The formation of wood in forest trees,* ed. M. H. Zimmermann, pp. 389–404. New York: Academic Press. *8, 65, 66, 97*

Brown, C. L. (1970). Physiology of wood formation in conifers. *Wood Sci.* **3,** 8–22. *55, 57, 61*

Brown, C. L. & Sax, K. (1962). The influence of pressure on the differentiation of secondary tissues. *Am. J. Bot.* **49,** 683–91. *8, 65, 97*

Brown, C. L. & Wodzicki, T. J. (1969). A simple technique for investigating cambial activity and the differentiation of cambial derivatives. *Forest Sci.* **15,** 26–9. *57*

Brown, D. D. & Dawid, I. B. (1969). Developmental genetics. *Annu. Rev. Genetics* **3,** 127–54. *42*

Brown, M. R., Franke, W. W. Kleinig, H., Falk, H. & Sitte, P. (1970). Scale formation in chrysophaceae algae. I. Cellulosic and noncellulosic wall components made by the Golgi apparatus. *J. Cell Biol.* **45,** 246–71. *79*

Brown, S. A. (1966). Lignins. *Annu. Rev. Plant Physiol.* **17,** 223–44. *82*

Brown, S. A. (1969). Biochemistry of lignin formation. *Bioscience* **19,** 115–21. *82*

Brunori, A. (1971). Synthesis of DNA and mitosis in relation to cell differentiation in the roots of *Vicia faba* and *Lactuca sativa. Caryologia* **24,** 209–15. *52*

Brunori, A. & D'Amato, F. (1967). The DNA content of nuclei in the embryo of dry seeds of *Pinus pinea* and *Lactuca sativa. Caryologia* **20,** 153–61. *51*

Bücher, T. & Sies, H. (eds.) (1969). *Inhibitors – tools in cell research.* Berlin: Springer-Verlag. *106*

Bullough, W. S. (1972). Epidermal chalone mechanism. *Proc. First Int. Chalone Conference,* ed. B. Forscher. Bethesda: National Cancer Institute (in press). *44*

Burg, S. P. & Burg, E. A. (1967). Molecular requirements for the biological activity of ethylene. *Plant Physiol.* **42,** 144–52. *103, 106*

Burgess, J. & Northcote, D. H. (1968). The relationship between the endoplasmic reticulum and microtubular aggregation and disaggregation. *Planta* **80,** 1–14. *76*

Cahn, R. D. (1968). Factors affecting inheritance and expression of differentiation: some methods of analysis. In: *The stability of the differentiated state,* ed. H. Ursprung, pp. 58–84. Berlin: Springer-Verlag. *52*

Cameron, I. L. & Jeter, J. R., Jr. (1971). Relationship between cell proliferation and cytodifferentiation in embryonic chick tissues. In: *Developmental aspects of the cell cycle,* ed. I. L. Cameron, G. M. Padilla & A. M. Zimmermann, pp. 191–222. New York: Academic Press. *45, 46*

Camus, G. (1949). Recherches sur le rôle des bourgeons dans les phénomènes de morphogénèse. *Rev. Cytol. et Biol. Veg.* **9,** 1–199. *7*

Caruso, J. L. & Cutter, E. G. (1970). Morphogenetic aspects of a leafless mutant in tomato. II. Induction of a vascular cambium. *Am. J. Bot.* **57,** 420–9. *68*

Catesson, A. M. (1974). Cambial cells. In: *Dynamic aspects of plant ultrastructure,* ed. A. W. Robards, pp. 358–90. New York: McGraw-Hill. *77, 78*

Catesson, A. M. & Czaninski, Y. (1968). Localisation ultrastructurale de la phosphatase acide et cycle saisonnier dans les tissus conducteurs de quelques arbres. *Bull. Soc. Franç. Physiol. Veg.* **14,** 165–73. *85*

Cawthon, G. E. (1972). The role of cytokinin and factors affecting xylem differentiation in Romaine lettuce pith cultured *in vitro.* PhD Thesis, University of Idaho, Moscow, USA. *24, 28, 91, 92*

Chafe, S. C. & Durzan, D. J. (1973). Tannin inclusions in cell suspension cultures of white spruce. *Planta* **113,** 251–62. *30*

Chafe, S. C. & Wardrop, A. B. (1970). Microfibril orientation in plant cell walls. *Planta* **92,** 13–24. *74*

Christie, A. E. & Leopold, A. C. (1965). On the manner of triiodobenzoic acid inhibition of auxin transport. *Plant Cell Physiol.* **6,** 337–45. *107*

Clutter, M. E. (1960). Hormonal induction of vascular tissue in tobacco pith *in vitro. Science* **132,** 548–9. *7*

Cook, P. R. (1974). On the inheritances of differentiated traits. *Biol. Rev.* **49,** 51–84. *41*

Cooper, H. L. (1971). Biochemical alterations accompanying initiation of growth in resting cells In: *The cell cycle and cancer,* ed. R. Baserga, pp. 197–226. New York: Marcel Dekker. *39, 40*

Cooper, W. C. (1972). Trauma-induced ethylene production by citrus flowers, fruit, and wood. In: *Plant growth substances 1970,* ed. D. J. Carr, pp. 543–8. New York: Springer-Verlag. *32, 101*

Corcoran, M. R., Geissman, T. A. & Phinney, B. O. (1972). Tannins as gibberellin antagonists. *Plant Physiol.* **49,** 323–30. *30*

Crocker, E. C. (1921). An experimental study of the significance of 'lignin' color reactions. *Ind. Eng. Chem.* **13,** 625–7. *82*

Cronshaw, J. (1965*a*). The organization of cytoplasmic components during the phase of cell wall thickening in differentiating cambial derivatives of *Acer rubrum. Can. J. Bot.* **43,** 1401–7. *71, 77, 81*

Cronshaw, J. (1965*b*). Cytoplasmic fine structure and cell wall development in differentiating xylem elements. In: *Cellular ultrastructure of woody plants,* ed. W. A. Côté, Jr., pp. 99–124. Syracuse: Syracuse University Press. *71, 81*

Cronshaw, J. (1967). Tracheid differentiation in tobacco pith cultures. *Planta* **72,** 78–90. *74, 81*

Crüger, H. (1855). Zur Entwicklungsgeschichte der Zellwand. *Bot. Ztg.* **13,** 601–13; 617–29. *4*

Cumming, D. F. (1970). Separation and identification of soluble nucleotides in cambial and young xylem tissue of *Larix decidua. Biochem. J.* **116,** 189–98. *94*

Cummings, B. G. (1959). The control of growth and development in red clover (*Trifolium pratense* L.). *Can. J. Bot.* **37,** 1049–54. *107*

Cutter, E. G. & Feldman, L. J. (1970). Trichoblasts in *Hydrocharis.* II. Nucleic acids, proteins and a consideration of cell growth in relation to endopolyploidy. *Am. J. Bot.* **57,** 202–11. *49, 50*

Czaninski, Y. (1968). Étude du parenchyme ligneux du Robinier (parenchyme à réserves et cellules associées aux vaisseaux) au cours du cycle annuel. *J. Microscopie* **7,** 145–64. *77, 79*

Czaninski, Y. (1970). Étude cytologique de la différenciation du bois de Robinier. II. Différenciation des cellules du parenchyme (cellules à réserves et cellules associées aux vaisseaux). *J. Microscopie* **9,** 389–406. *77, 79*

Czaninski, Y. (1972*a*). Mise en évidence de cellules associées aux vaisseaux dans le xylème du Sycomore. *J. Microscopie* **13,** 137–40. *79*

Czaninski, Y. (1972*b*). Observations ultrastructurales sur l'hydrolyse des parois primaires des vaisseaux chez le *Robinia pseudo-acadia* L. et l'*Acer pseudoplatanus* L. *C. R. Acad. Sci.* (Paris) **275,** 361–3. *77*

Czaninski, Y. (1973). Observations on a new wall layer in vessel associated cells of *Robinia* and *Acer*. *Protoplasma* **77,** 211–19. *79*

Czaninski, Y. & Catesson, A. M. (1969). Ultrastructural localization of peroxidase activity in plant conductive tissues during the annual cycle. *J. Microscopie* **8,** 875–88. *88*

Dalessandro, G. (1973*a*). Hormonal control of xylogenesis in pith parenchyma explants of *Lactuca*. *Ann. Bot.* **37,** 375–82. *2, 7, 22*

Dalessandro, G. (1973*b*). Interaction of auxin, cytokinin, and gibberellin on cell division and xylem differentiation in cultured explants of Jerusalem artichoke. *Plant Cell Physiol.* **14,** 1167–76. *16, 17, 21, 25, 27, 29, 34, 49, 62, 107*

Dalessandro, G. & Roberts, L. W. (1971). Induction of xylogenesis in pith parenchyma explants of *Lactuca*. *Am. J. Bot.* **58,** 378–85. *7, 16, 17, 18, 19, 20–2, 45, 92, 109*

D'Amato, F. (1952*a*). Polyploidy in the diffentiation and function of tissues and cells in plants. *Caryologia* **4,** 311–58. *49*

D'Amato, F. (1952*b*). New evidence on endopolyploidy in differentiated plant tissues. *Caryologia* **4,** 121–44. *49*

D'Amato, F. (1953). Studio anatomico delle reazioni indotte dalla decapitazione combinata a 2,4-D in turioni di *Asparagus officinalis* L. *Nuovo G. Bot. Ital.* **60,** 660–79. *45*

D'Amato, F. (1960). Cyto-histological investigations of antimitotic substances and their effects on patterns of differentiation. *Caryologia* **13,** 339–51. *106*

D'Amato, F. (1965). Endopolyploidy as a factor in plant tissue development. In: *Proc. int. conf. on plant tissue culture,* ed. P. R. White & A. R. Grove, pp. 449–62. Berkeley: McCutchan. *49, 50*

D'Amato, F. & Avanzi, M. G. (1948). Reazioni di natura auxinica et effecti rizogeni in *Allium cepa* L. Studio cito-istologico sperimentale. *Nuovo G. Bot. Ital.* **55,** 161–213. *49*

Davies, L. M., Priest, J. & Priest, R. (1968). Collagen synthesis by cell synchronously replicating DNA. *Science* **159,** 91. *46*

Davies, P. J. (1973). Current theories on the mode of action of auxin. *Bot. Rev.* **39,** 139–71. *12, 13*

Davis, J. D. & Evert, R. F. (1965). Phloem development in *Populus tremuloides*. *Am. J. Bot.* **52,** 627. *62*

Dawes, I., Kay, D. & Mandelstam, J. (1971). Determining effect of growth medium on the shape and position of daughter chromosomes and on sporulation in *Bacillus subtilis*. *Nature* (Lond.) **230,** 567–9. *41*

Degani, Y. & Atsmon, D. (1970). Enhancement of non-nuclear DNA synthesis associated with hormone-induced elongation in the cucumber hypocotyl. *Exp. Cell Res.* **61,** 226–9. *48*

Degani, Y., Atsmon, D. & Halevy, A. (1970). DNA synthesis and hormone-induced elongation in cucumber hypocotyl. *Nature* (Lond.) **228,** 554–5. *48*

De Jong, D. W. (1967). An investigation of the role of plant peroxidase in cell wall development by the histochemical method. *J. Histochem. Cytochem.* **15,** 335–46. *88*

Delmer, D. P. & Albersheim, P. (1970). The biosynthesis of sucrose and nucleoside diphosphate glucoses in *Phaseolus aureus. Plant Physiol.* **45,** 782–6. *92, 93*

DeMaggio, A. E. (1966). Phloem differentiation: induced stimulation by gibberellic acid. *Science* **152,** 370–2. *114*

DeMaggio, A. E. (1972). Induced vascular tissue differentiation in fern gametophytes. *Bot. Gaz.* **133,** 311–17. *20, 49, 91, 92*

Denne, M. P. (1970). Xylem development in conifers. In: *Physiology of tree crops,* ed. L. C. Luckwill & C. V. Cutting, pp. 83–96. New York: Academic Press. *55*

Denne, M. P. (1971). Temperature and tracheid development in *Pinus sylvestris* seedlings. *J. Exp. Bot.* **22,** 362–70. *99*

Deysson, G. (1968). Antimitotic substances. *Int. Rev. Cytol.* **24,** 99–148. *106*

Digby, J. & Wareing, P. F. (1966). The relationship between endogenous hormone levels in the plant and seasonal aspects of cambial activity. *Ann. Bot.* **30,** 607–22. *8, 56, 58, 61*

Dippel, L. (1867). Die Entstehung der wandständigen Protoplasmaströmchen. *Abh. Naturforsch. Ges. Halle* **10,** 53–68. *4*

Dobbins, D. R. (1971). Studies on the anomalous cambial activity in *Doxantha unguis-cati* (Bignoniaceae). II. A case of differential production of secondary tissues. *Am. J. Bot.* **58,** 697–705. *111, 113*

Doley, D. (1970). Effects of simulated drought on shoot development in *Liriodendron* seedlings. *New Phytol.* **69,** 655–73. *95, 96, 97*

Doley, D. & Leyton, L. (1970). Effects of growth regulating substances and water potential on the development of wound callus in *Fraxinus. New Phytol.* **69,** 87–102. *7, 28, 96, 97*

Donachie, W. D. & Masters, M. (1969). Temporal control of gene expression in bacteria. In: *The cell cycle, gene-enzyme interactions,* ed. G. M. Padilla, G. L. Whitson & I. L. Cameron, pp. 37–76. New York: Academic Press. *44*

Duffy, S. L. (1965). The effects of boron, copper, and manganese on xylem differentiation around wounds in stems of *Coleus blumei* Benth. Thesis, Howard University, Washington, DC. *89*

Dunlop, D. W. & Schmidt, B. L. (1964). Biomagnetics, I. Anomalous development of the root of *Narcissus tazetta* L. *Phytomorphology* **14,** 333–42. *102*

Dunlop, D. W. & Schmidt, B. L. (1965). Biomagnetics. II. Anomalies found in the root of *Allium cepa* L. *Phytomorphology* **15,** 400–14. *102*

Dunlop, D. W. & Schmidt, B. L. (1969). Sensitivity of some plant material to magnetic fields. In: *Biological effects of magnetic fields,* vol. 2, ed. M. F. Barnothy, pp. 147–70. New York: Plenum Press. *102*

Durzan, D. J., Chafe, S. C. & Lopushanski, S. M. (1973). Effects of environmental changes on sugars, tannins, and organized growth in cell suspension cultures of white spruce. *Planta* **113,** 214–9. *30, 98*

Ebert, J. D. & Sussex, I. M. (1970). *Interacting systems in development,* 2nd edition. New York: Holt, Rinehart & Winston. *41, 51*

Eigsti, O. J. (1938). A cytological study of colchicine effects in the induction of polyploidy in plants. *Proc. Nat. Acad. Sci* (USA) **24,** 56–63. *72*

Engelke, A. L., Hamzi, H. Q. & Skoog, F. (1973). Cytokinin–gibberellin regulation of shoot development and leaf form in tobacco plantlets. *Am. J. Bot.* **60,** 491–5. *15, 29*

Ephrussi, B. (1972). *Hybridization of somatic cells.* Princeton: Princeton Uinversity Press. *2, 41, 42*

Ernst, R., Arditti, J. & Healey, P. L. (1971). Carbohydrate physiology of orchid seedlings. II. Hydrolysis and effects of oligosaccharides. *Am. J. Bot.* **58,** 827–35. *91*

Esau, K. (1965*a*). *Plant anatomy*, 2nd edition. New York: John Wiley. *1, 54, 101*

Esau, K. (1965*b*). *Vascular differentiation in plants*. New York: Holt, Rinehart & Winston. *3, 10, 114*

Esau, K., Cheadle, V. I. & Gill, R. H. (1966*a*). Cytology of differentiating tracheary elements. I. Organelles and membrane systems. *Am. J. Bot.* **53,** 756–64. *70, 71*

Esau, K., Cheadle, V. I. & Gill, R. H. (1966*b*). Cytology of differentiating tracheary elements. II. Structures associated with cell surfaces. *Am. J. Bot.* **53,** 765–71. *70, 71*

Eschrich, W. (1968). Translokation radioaktiv markierter Indolyl-3-Essigsäure in Siebrohren von *Vicia faba*. *Planta* **78,** 144–78. *61*

Evans, L. S. & Van't Hof, J. (1973). Cell arrest in G_2 in root meristems. A control factor from the cotyledons. *Exp. Cell Res.* **82,** 471–3. *44*

Evans, M. L. (1974). Rapid responses to plant hormones, *Annu. Rev. Plant Physiol.* **25,** 195–223. *13*

Evert, R. F. (1960). Phloem structure in *Pyrus communis* L. and its seasonal changes. *Univ. Calif. Pub. Bot.* **32,** 127–94. *62*

Evert, R. F. (1963). The cambium and seasonal development of the phloem in *Pyrus malus*. *Am. J. Bot.* **50,** 149–59. *62*

Evert, R. F. & Kozlowski, T. T. (1967). Effect of isolation of bark on cambial activity and development of xylem and phloem in trembling aspen. *Am. J. Bot.* **54,** 1045–54. *8, 57, 58*

Evert, R. F., Kozlowski, T. T. & Davis, J. D. (1972). Influence of phloem blockage on cambial growth of sugar maple. *Am. J. Bot.* **59,** 632–41. *8, 58*

Evins, W. H. & Varner, J. E. (1971). Hormone-controlled synthesis of endoplasmic reticulum in barley aleurone cells. *Proc. Nat. Acad. Sci.* (USA) **68,** 1631. *29*

Farber, E. & Baserga, R. (1969). Differential effects of hydroxyurea on survival of proliferating cells. *Cancer Res.* **29,** 136–9. *40*

Ferré, S. R. (1971). The role of kinetin and adenine sulfate in lignification and xylogenesis of callus tissue of *Parthenocissus tricuspidata*. MS Thesis, Rutgers University, New Brunswick, NJ. *8, 35*

Field, R. J. (1973). The effect of gibberellic acid, kinetin and indolylacetic acid on the growth regulators in willow stems. *New Phytol.* **72,** 471–8. *62*

Foard, D. E. (1960). An experimental study of sclereid formation in *Camellia japonica*. PhD Thesis, North Carolina State University, Raleigh. *45, 112*

Foard, D. E. (1970). Differentiation in plant cells. In: *Cell differentiation*, ed. O. A. Schjeide & J. De Vellis, pp. 575–602. New York: Van Nostrand Reinhold. *45, 112*

Foard, D. E. & Haber, A. H. (1961). Anatomic studies of gamma-irradiated wheat growing without cell division. *Am. J. Bot.* **48,** 438–46. *45, 47*

Forest, J. C. & McCully, M. E. (1971). Histological study on the *in vitro* induction of vascularization in tobacco pith parenchyma. *Can. J. Bot.* **49,** 449–52. *18*

Fosket, D. E. (1968). Cell division and the differentiation of wound-vessel members in cultured stem segments of *Coleus*. *Proc. Nat. Acad. Sci.* (USA) **59,** 1089–96. *22, 38*

Fosket, D. E. (1970). The time course of xylem differentiation and its relation to deoxyribonucleic acid synthesis in cultured *Coleus* stem segments. *Plant Physiol.* **46,** 64–8. *13, 14, 38, 41, 48*

Fosket, D. E. (1972). Meristematic activity in relation to wound xylem differentiation. In: *The dynamics of meristem cell populations*, ed. M. W. Miller & C. C. Kuehnert, pp. 33–50. New York: Plenum Press. *38, 48*

Fosket, D. E. & Miksche, J. P. (1966). Protein synthesis as a requirement for wound xylem differentiation. *Physiol. Plant.* **19,** 982–91. *15, 84*

Fosket, D. E. & Roberts, L. W. (1964). Induction of wound-xylem differentiation in isolated *Coleus* stem segments *in vitro*. *Am. J. Bot.* **51,** 19–25. *7, 20, 89*

Fosket, D. E. & Short, K. C. (1973). The role of cytokinin in the regulation of growth, DNA synthesis and cell proliferation in cultured soybean tissues (*Glycine max* var. Biloxi). *Physiol. Plant.* **28,** 14–23. *24, 48, 49*

Fosket, D. E. & Torrey, J. G. (1969). Hormonal control of cell proliferation and xylem differentiation in cultured tissues of *Glycine max* var. Biloxi. *Plant Physiol.* **44,** 871–80. *7, 20–3, 35, 38*

Foster, A. S. (1944). Structure and development of sclereids in the petiole of *Camellia japonica*. *Bull. Torrey Bot. Club* **71,** 302–26. *45*

Fowke, L. C. & Pickett-Heaps, J. D. (1972). A cytochemical and autoradiographic investigation of cell wall deposition in fiber cells of *Marchantia berteroana*. *Protoplasma* **74,** 19–32. *76*

Fragata, M. (1970). Mitotic apparatus. A possible site of action of gibberellic acid. *Naturwiss.* **57,** 139. *29*

Frankfurt, O. S. (1971). Epidermal chalone. Effect on cell cycle of development of hyperplasia. *Exp. Cell Res.* **64,** 140–4. *44*

Freudenberg, K. (1964). The formation of lignin in the tissue and *in vitro*. In: *The formation of wood in forest trees,* ed. M. H. Zimmermann, pp. 203–18. New York: Academic Press. *82*

Freudenberg, K. (1965). Lignin: its constitution and formation from *p*-hydroxycinnamyl alcohols. *Science* **148,** 595–600. *82*

Freundlich, H. F. (1908). Entwicklung und Regeneration von Gefässbündeln in Blattgebilden. *Jahrb. wiss. Bot.* **46,** 137–206. *4*

Fritsch, F. E. (1945). *The structure and reproduction of the algae,* vol. 2, pp. 232–5. London: Cambridge University Press. *114*

Gagnon, C. (1968). Peroxidase in healthy and diseased elm trees investigated by the benzidine histochemical technique. *Can. J. Bot.* **46,** 1491–4. *88*

Gahan, P. B. (1973). Plant lysosomes. In: *Lysosomes in biology and pathology,* vol. 3, ed. J. T. Dingle, pp. 69–85. Amsterdam: North Holland Publishing Co. Ltd. *76*

Gahan, P. B. & Maple, A. J. (1966). The behaviour of lysosome-like particles during cell differentiation. *J. Exp. Bot.* **17,** 151–5. *76, 85*

Gamaley, Ju. V. (1971). Autolysis in the differentiating tracheids. *Cytology* (Leningrad) **13,** 278–86. *77*

Gamaley, Ju. V. (1972). *Cytological bases of xylem differentiation*. Leningrad: Botanical Institute, Acad. Sci. USSR. *3*

Gamborg, O. L. & LaRue, T. A. G. (1971). Ethylene production by plant cell cultures. The effect of auxins, abscisic acid and kinetin on ethylene production in suspension cultures or rose and *Ruta* cells. *Plant Physiol.* **48,** 399–401. *103*

Gaspar, T., Khan, A. A. & Fries, D. (1973). Hormonal control of isoperoxidases in lentil embryonic axis. *Plant Physiol.* **51,** 146–9. *86*

Gautheret, R. J. (1942). Le bourgeonnement des tissus végétaux en culture. *Sciences* (Paris) **40,** 95–128. *7*

Gautheret, R. J. (1961*a*). Action conjuguée de l'acide gibbérellique, de la cinétine et de l'acid indole-acétique sur les tissus cultivés *in vitro,* particulièrement sur ceux de Topinambour. *C.R. Acad. Sci.* (Paris) **253,** 1381–5. *7, 25*

Gautheret, R. J. (1961*b*). Action de la lumière et de la température sur la néoformation de racines par des tissus de Topinambour cultivés *in vitro*. *C.R. Acad. Sci.* (Paris) **250,** 2791–6. *31, 99, 100*

Gautheret, R. J. (1966). Factors affecting differentiation of plant tissues grown *in vitro*. In: *Cell differentiation and morphogenesis,* ed. W. Beermann, pp. 55–95. Amsterdam: North-Holland Publishing Co. Ltd. *1, 8, 64*

Gee, H. (1972). Localization and uptake of ^{14}C-IAA in relation to xylem regeneration in *Coleus* internodes. *Planta* **108,** 1–9. *45, 85*

Gelfant, S. (1963). Inhibition of cell division: a critical and experimental analysis. *Int. Rev. Cytol.* **14,** 1–39. *106*

Ghorashy, S. R. (1967). Effects of 2,3,5-triiodobenzoic acid on the anatomy of *Glycine max* (L.) Merril. PhD. Thesis, University of Nebraska, Lincoln. *108*

Goeschl, J. D., Rappaport, L. & Pratt, H. K. (1966). Ethylene as a factor regulating the growth of pea epicotyls subjected to physical stress. *Plant Physiol.* **41,** 877–89. *66*

Goldstein, M. A., Claycomb, W. C. & Schwartz, A. (1974). DNA synthesis and mitosis in well-differentiated mammalian cardiocytes. *Science* **183,** 212–13. *46*

González-Fernández, A., Díez, J. L., Giménez-Martín, G. & De La Torre, C. (1972). Direct measurement of interphase shortening produced by kinetin plus indoleacetic acid in meristematic cells of *Allium cepa* L. *Experientia* **28,** 247–8. *47*

Goosen-De Roo, L. (1973*a*). The fine structure of the protoplast in primary tracheary elements of the cucumber after plasmolysis. *Acta Bot. Neerl.* **22,** 467–85. *4, 71*

Goosen-De Roo, L. (1973*b*). The relationship between cell organelles and cell wall thickenings in primary tracheary elements of the cucumber. I. Morphological aspects. *Acta. Bot. Neerl.* **22,** 279–300. *4*

Goosen-De Roo, L. (1973*c*). The relationship between cell organelles and cell wall thickenings in primary tracheary elements of the cucumber. II. Quantitative aspects. *Acta Bot. Neerl.* **22,** 301–20. *4*

Granick, S. & Gibor, A. (1967). The DNA of chloroplasts, mitochondria, and centrioles. In: *Progress in nucleic acid research and molecular biology,* vol. 6, ed. J. N. Davidson & W. E. Cohn, pp. 143–86. New York: Academic Press. *52*

Green, H. & Todaro, G. J. (1967). The mammalian cell as a differentiated microorganism. *Annu. Rev. Microbiol.* **21,** 573–600. *41*

Green, P. (1969). Cell morphogenesis. *Annu. Rev. Plant Physiol.* **20,** 365–94. 29

Greenway, H., Hiller, R. G. & Flowers, T. J. (1968). Respiratory inhibition in *Chlorella* produced by 'purified' polyethylene glycol 1540. *Science* **159,** 984–5. *97*

Gross, P. R. (1968). Biochemistry of differentiation. *Annu. Rev. Biochem.* **37,** 631–60. *1, 41, 42*

Guttman, R. (1956). Effects of kinetin on cell division, with special reference to initiation and duration of mitosis. *Chromosoma* **8,** 341–50. *47*

Haber, A. H. & Foard, D. E. (1964). Further studies on gamma-irradiated wheat and their relevance to use of mitotic inhibition for developmental studies. *Am. J. Bot.* **51,** 151–9. *45, 47*

Haber, A. H. & Schwarz, O. J. (1972). A method for testing the specificity of inhibitors of deoxyribonucleic acid synthesis in growth studies. *Plant Physiol.* **49,** 335–7. *40*

Hall, M. A. & Ordin, L. (1968). Auxin-induced control of cellulase synthetase activity in *Avena* coleoptile sections. In: *Biochemistry and physiology of plant growth substances,* ed. F. Wightman & G. Setterfield, pp. 659–71. Ottawa: Runge Press. *16*

Hall, R. H. (1973). Cytokinins as a probe of developmental processes. *Annu. Rev. Plant Physiol.* **24,** 415–44. *107*

Halperin, W. (1969). Morphogenesis in cell cultures. *Annu. Rev. Plant Physiol.* **20,** 395–418. *3, 8, 28*

Halvorson, H. O., Bock, R. M., Tauro, P., Epstein, R. L. & La Berge, M. (1966). Periodic enzyme synthesis in synchronous cultures of yeast. In: *Cell synchrony. Studies in biosynthetic regulation,* ed. I. L. Cameron & G. M. Padilla, pp. 102–16. New York: Academic Press. *44*

Halvorson, H. O., Carter, B. L. A. & Tauro, P. (1971). Synthesis of enzymes during the cell cycle. *Adv. Microbial Physiol.* **6,** 47–106. *44*

Harkin, J. M. & Obst, J. R. (1973). Lignification in trees: indication of exclusive peroxidase participation. *Science* **180,** 296–8. *87*

Hartig, T. (1853). Uber die Entwicklung des Jahrringes der Holzpflanzen. *Bot. Ztg.* **11,** 553–60; 569–79. *8*

Hawker, J. S. (1971). Enzymes concerned with sucrose synthesis and transformations in seeds of maize, broad bean and castor bean. *Phytochem.* **10,** 2313–22. *90*

Hébant, Ch. (1973). Acid phosphomonoesterase activities (β-glycerophosphatase and naphthol AS-MX phosphatase) in conducting tissues of bryophytes. *Protoplasma* **77,** 231–41. *85*

Hecht, S. M., Bock, R. M., Schmitz, R. Y., Skoog, F. & Leonard, N. J. (1971. Cytokinins: development of a potent antagonist. *Proc. Nat. Acad. Sci.* (USA) **68,** 2608–10. *25*

Hejnowicz, A. & Tomaszewski, M. (1969). Growth regulators and wood formation in *Pinus silvestris. Physiol. Plant.* **22,** 984–92. *61*

Hepler, P. K. & Fosket, D. E. (1971). The role of microtubules in vessel member differentiation in *Coleus. Protoplasma* **72,** 213–36. *71–3, 79*

Hepler, P. K., Fosket, D. E. & Newcomb, E. H. (1970). Lignification during secondary wall formation in *Coleus:* an electron microscopic study. *Am. J. Bot.* **57,** 85–96. *71, 79, 81–4, 87*

Hepler, P. K. & Newcomb, E. H. (1963). The fine structure of young tracheary xylem elements arising by redifferentiation of parenchyma in wounded *Coleus* stem. *J. Exp. Bot.* **14,** 496–503. *79*

Hepler, P. K. & Newcomb, E. H. (1964). Microtubules and fibrils in the cytoplasm of *Coleus* cells undergoing secondary wall deposition. *J. Cell Biol.* **20,** 529–33. *79*

Hepler, P. K. & Palevitz, B. A. (1974). Microtubules and microfilaments. *Annu. Rev. Plant Physiol.* **25,** 309–62. *71, 74*

Hepler, P. K., Rice, R. M. & Terranova, W. A. (1972). Cytochemical localization of peroxidase activity in wound vessel members of *Coleus. Can. J. Bot.* **50,** 977–83. *74, 79, 81–2, 84, 86–8*

Heslop-Harrison, J. (1967). Differentiation. *Annu. Rev. Plant Physiol.* **18,** 325–48. *1, 12, 45*

Hess, T. & Sachs, T. (1972). The influence of a mature leaf on xylem differentiation. *New Phytol.* **71,** 903–14. *8, 36, 62, 63*

Higuchi, T. (1957). Biochemical studies of lignin formation. I, II, III. *Physiol. Plant.* **10,** 346–72; 621–32; 633–48. *88*

Hoad, G. V., Hillman, S. K. & Wareing, P. F. (1971). Studies on the movement of indole auxins in willow (*Salix viminalis* L.). *Planta* **99,** 73–88. *61*

Hochster, R. M. & Quastel, J. H. (eds) (1963). *Metabolic inhibitors. A comprehensive treatise.* 2 vols. New York: Academic Press. *106*

Hogetsu, T., Shibaoka, H. & Shimokoriyama, M. (1974*a*). Involvement of cellulose synthesis in actions of gibberellin and kinetin on cell expansion. Gibberellin–coumarin and kinetin–coumarin interactions on stem elongation. *Plant Cell Physiol.* **15,** 265–72. *25, 29*

Hogetsu, T., Shibaoka, H. & Shimokoriyama, M. (1974*b*). Involvement of cellulose synthesis in actions of gibberellin and kinetin on cell expansion. 2,6-dichlorobenzonitrile as a new cellulose-synthesis inhibitor. *Plant Cell Physiol.* **15,** 389–93. *25, 29*

Holtzer, H. (1970). Proliferative and quantal mitosis in differentiation. In: *Control mechanisms in the expression of cellular phenotypes,* ed. H. A. Padykula, pp. 69–88. New York: Academic Press. *38, 39*

Holtzer, H. & Abbott, J. (1968). Oscillations of chondrogenic phenotypes *in vitro.* In:

The stability of the differentiated state, ed. H. Ursprung, pp. 1–16. Berlin: Springer-Verlag. *1, 2, 52*

Holtzer, H. & Bischoff, R. (1970). Mitosis and myogenesis. In: *The physiology and biochemistry of muscle as a food*, vol. 2, ed. E. J. Briskey, R. G. Cassens & B. B. Marsh, pp. 29–51. Madison: University of Wisconsin Press. *1, 41*

Hook, D. D. & Brown, C. L. (1972). Permeability of the cambium to air in trees adapted to wet habitats. *Bot. Gaz.* **133,** 304–10. *95*

Howard, A. & Pelc, S. (1953). Synthesis of deoxyribonucleic acid in normal and irradiated cells and its relation to chromosome breakage. *Heredity* **6** (suppl.), 261–73. *38*

Hsiao, T. C. (1973). Plant responses to water stress. *Annu. Rev. Plant Physiol.* **24,** 519–70. *98*

Hummon, M. R. (1962). The effects of tritiated thymidine incorporation on secondary root production by *Pisum sativum*. *Am. J. Bot.* **49,** 1038–46. *72*

Innocenti, A. M. (1973). Élaboration des parois des jeunes cellules métaxylémiennes dans la racine de l'*Allium cepa*, cultivar *fiorentina*. Application de la méthode de Théiry et localisation de l'activité peroxydasique. *C.R. Acad. Sci.* (Paris) **277,** 2677–80. *86, 88*

Innocenti, A. M. & Avanzi, S. (1971). Some cytological aspects of the differentiation of metaxylem in the root of *Allium cepa*. *Caryologia* **24,** 283–91. *13, 51*

Isherwood, F. A. (1965). Biosynthesis of lignin. In: *Biosynthetic pathways in higher plants*, ed. J. B. Pridham & T. Swain, pp. 133–46. New York: Academic Press. *82*

Itoh, T. (1971). On the ultrastructure of dormant and active cambium of conifers. *Wood Res.* (Kyoto Univ.) **51,** 33–5. *78*

Iwamura, T. (1966). Nucleic acids in chloroplasts and metabolic DNA. In: *Progress in nucleic acid research and molecular biology*, vol. 5, ed. J. N. Davidson & W. E. Cohn, pp. 133–55. New York: Academic Press. *52*

Jackson, W. T. (1962). Use of carbowaxes (polyethylene glycols) as osmotic agents. *Plant Physiol.* **37,** 513–19. *97*

Jacob, F. & Monod, J. (1963). Genetic repression, allosteric inhibition and cellular differentiation. In: *Cytodifferentiation and macromolecular synthesis*, ed. M. Locke, pp. 30–64. New York: Academic Press. *1, 89*

Jacobs, W. P. (1952). The role of auxin in differentiation of xylem around a wound. *Am. J. Bot.* **39,** 301–9. *4, 12*

Jacobs, W. P. (1954). Acropetal auxin transport and xylem regeneration – a quantitative study. *Am. Nat.* **88,** 327–37. *4, 12*

Jacobs, W. P. (1956). Internal factors controlling cell differentiation in the flowering plants. *Am. Nat.* **90,** 163–9. *4*

Jacobs, W. P. (1959). What substance normally controls a given biological process? I. Formulation of some rules. *Develop. Biol.* **1,** 527–33. *4, 5*

Jacobs, W. P. (1970*a*). Regeneration and differentiation of sieve tube elements. *Int. Rev. Cytol.* **28,** 239–73. *114*

Jacobs, W. P. (1970*b*). Development and regeneration of the algal giant coenocyte *Caulerpa*. *Ann. N.Y. Acad. Sci.* **175,** 712–31. *114*

Jacobs, W. P. & Morrow, I. B. (1957). A quantitative study of xylem development in the vegetative shoot apex of *Coleus*. *Am. J. Bot.* **44,** 823–42. *4, 12*

Jacquoit, C. (1947). Effet inhibiteur des tannins sur le développement des cultures *in vitro* du cambium de certains arbres forestiers. *C.R. Acad. Sci.* (Paris) **225,** 434–6. *31*

Jacquoit, C. (1950). Contribution a l'étude des facteurs determinant de cycle d'activité du cambium chez quelques arbres forestiers. *Rev. Forst. Franc.* **2,** 605–10. *60*

Jacquiot, C. (1957). Sur l'existence de facteurs d'organisation des tissus secondaires chez certains arbres forestiers. *C.R. Acad. Sci.* (Paris) **244,** 1246–8. *64*

Jeffs, R. A. & Northcote, D. H. (1966). Experimental induction of vascular tissue in an undifferentiated plant callus. *Biochem. J.* **101,** 146–52. *93*

Jeffs, R. A. & Northcote, D. H. (1967). The influence of indol-3yl-acetic acid and sugar on the pattern of induced differentiation in plant tissue cultures. *J. Cell Sci.* **2,** 77–88. *7, 89, 90–2, 94*

Jensen, W. A. (1955). The histochemical localization of peroxidase in roots and its induction by indoleacetic acid. *Plant Physiol.* **30,** 426–32. *88*

Jocelyn, P. C. (1972). *Biochemistry of the SH group; the occurrence, chemical properties, metabolism and biological function of thiols and disulfides.* New York: Academic Press. *110*

Jones, D. T. & Villiers, T. A. (1972). Changes in distribution of acid phosphatase activity during wound regeneration in *Coleus. J. Exp. Bot.* **23,** 375–80. *85*

Jones, R. L. (1973). Gibberellins: their physiological role. *Annu. Rev. Plant Physiol.* **24,** 571–98. *31*

Jost, L. (1940). Zur Physiologie der Gefässbildung. *Z. Bot.* **35,** 114–50. *4*

Jost, L. (1942). Uber Gefässbrucken. *Z. Bot.* **38,** 161–215. *4*

Jouanneau, J. P. (1971). Contrôle par les cytokinines de la synchronisation des mitoses dans les cellules de tabac. *Exp. Cell Res.* **67,** 329–37. *24, 48*

Jouanneau, J. P. & Tandeau de Marsac, N. (1973). Stepwise effects of cytokinin activity and DNA synthesis upon mitotic cycle events in partially synchronized tobacco cells. *Exp. Cell res.* **77,** 167–74. *48*

Kaan Albest, A. von. (1934). Anatomische und physiologische Untersuchungen über die Entstehung von Siebröhrenverbindungen. *Z. Bot.* **27,** 1–94. *4*

Karstens, W. H. K. & de Meester-Manger Cats, V. (1960). The cultivation of plant tissues *in vitro* with starch as a source of carbon. *Acta Bot. Neerl.* **9,** 263–74. *7*

Kaufman, P. B., Ghosheh, N. S., LaCroix, J. D., Soni, S. L. & Ikuma, H. (1973). Regulation of invertase levels in *Avena* stem segments by gibberellic acid, sucrose, glucose, and fructose. *Plant Physiol.* **52,** 221–8. *30, 90*

Keitt, G. W., Jr. & Baker, R. A. (1966). Auxin activity of substituted benzoic acids and their effect on polar auxin transport. *Plant Physiol.* **41,** 1561–9. *107*

Kennedy, R. W. (1970). An outlook for basic wood anatomy research. *Wood and Fiber* **2,** 182–7. *101*

Kennedy, R. W. & Farrar, J. L. (1965). Tracheid development in tilted seedlings. In: *Cellular ultrastructure of woody plants,* ed. W. A. Côté, Jr., pp. 419–53. Syracuse: Syracuse University Press. *63*

Kessell, R. H. J. & Carr, A. H. (1972). The effect of dissolved oxygen concentration on growth and differentiation of carrot (*Daucus carota*) tissue. *J. Exp. Bot.* **23,** 996–1007. *102*

Kessler, B. (1973). Hormonal and environmental modulation of gene expression in plant development. In: *The biochemistry of gene expression in higher organisms,* ed. J. K. Pollard & J. W. Lee, pp. 33–56. Artarmon: Australia & New Zealand Book Co. *35*

Key, J. L. & Vanderhoef, L. N. (1973). Plant hormones and developmental regulation: role of transcription and translation. In: *Developmental regulation: aspects of cell differentiation,* ed. S. J. Coward, pp. 49–83. New York: Academic Press. *12*

Kihlman, B. A. (1966). *Actions of chemicals on dividing cells.* Englewood Cliffs: Prentice-Hall. *106*

Kihlman, B. A., Eriksson, T. & Odmark, G. (1966). Effects of hydroxyurea on chromosomes, cell division and nucleic acid synthesis in *Vicia faba. Hereditas* (Lund) **55,** 386–97. *40*

Kirkham, M. B., Gardner, W. R. & Gerloff, G. C. (1972). Regulation of cell division and cell enlargement by turgor pressure. *Plant Physiol.* **49,** 961–2. *96*

Kitching, J. A. (1970). Some effects of high pressure on protozoa. In: *High pressure effects on cellular processes,* ed. A. M. Zimmerman, pp. 155–77. New York: Academic Press. *96*

Kleiber, H. & Mohr, H. (1967). Vom Einfluss des Phytochroms auf die Xylemdifferenzierung im Hypokotyl des Senfkimlings (*Sinapis alba* L.). *Planta* **76,** 85–92. *101*

Koblitz, H. (1961). Einige Bermerkungen zum histochemischen Nachweis von an der Verholzung beteiligten Fermenten. *Ber. Dtsch. Bot. Ges.* **74,** 274–87. *88*

Kochhar, V. K., Anand, V. K. & Nanda, K. K. (1972). Effect of morphactin on rooting and sprouting of buds on stem cuttings of *Salix tetrasperma. Bot. Gaz.* **133,** 361–8. *109*

Kollmann, R. & Schumacher, W. (1964). Über die Feinstruktur des phloems von *Metasequoia glyptostroboides* und seine jahreszietlichen veränderungen. V. Die Differenzierung der Siebzellen im Verlaufe einer Vegetationsperiode. *Planta* **63,** 155–90. *77*

Kozlowski, T. T. (1971). *Growth and development of trees,* vol. 2. *Cambial growth, root growth and reproductive growth.* New York: Academic Press. *3, 55*

Kratzl, K. (1965). Lignin – its biochemistry and structure. In: *Cellular ultrastructure of woody plants,* ed. W. A. Côté, Jr., pp. 157–80. Syracuse: Syracuse University Press. *82*

Krause, B. F. (1971). Structural and histological studies of the cambium and shoot meristems of soybeans treated with 2,3,5-triiodobenzoic acid. *Am. J. Bot.* **58,** 148–59. *108*

Krelle, E. (1970). Beiträge zur Physiologie der Morphaktinwirkung. III. Hemmung des Auxin-transports und Sulfhydrylaktivität. *Biochem. Physiol. Pflazen* **161,** 299–309. *110*

Kuo, J. & O'Brien, T. P. (1974). Lignified sieve elements in the wheat leaf. *Planta* **117,** 349–53. *114*

Lagerwerff, J. V., Ogata, G. & Eagle, H. E. (1961). Control of osmotic pressure of culture solutions with polyethylene. *Science* **133,** 1486. *97*

Lajtha, L. G. (1963). untitled. *J. Cell. Comp. Physiol.* (suppl. 1) 143–5. *39*

Lajtha, L. G. (1967). Proliferation kinetics of steady state cell populations. In: *Control of cellular growth in adult organisms,* ed. H. Teir & T. Rytomaa, pp. 97–105. New York: Academic Press. *39*

LaMotte, C. E. & Jacobs, W. P. (1963). A role of auxin in phloem regeneration in *Coleus* internodes. *Develop. Biol.* **8,** 80–98. *114*

LaMotte, C. E. & Houck, D. (1973). Studies of phloem regeneration unaccompanied by xylem regeneration in excised internodes: the requirement for both auxin and cytokinin. *Plant Physiol.* **51** (suppl.), 23. *114*

Landgren, C. R. & Torrey, J. G. (1973). The culture of protoplasts derived from explants of seedling pea roots. In: *Protoplastes et fusion de cellules somatiques végétales.* Colloques Internationaux CNRS no. 212, pp. 281–9. Paris:CNRS. *115*

LaRue, T. A. G. & Gamborg, O. L. (1971). Ethylene production by plant cell cultures. Variations in production during growing cycle and in different plant species. *Plant Physiol.* **48,** 394–8. *103*

Lasher, R. (1971). Studies on cellular proliferation and chondrogenesis. In: *Developmental aspects of the cell cycle,* ed. I. L. Cameron, G. M. Padilla & A. M. Zimmerman, pp. 223–41. New York: Academic Press. *52*

Lau, O. L. & Yang, S. F. (1973). Mechanism of a synergistic effect of kinetin on auxin-induced ethylene production. Suppression of auxin conjugation. *Plant Physiol.* **51,** 1011–14. *32*

Lavee, S. & Galston, A. W. (1968). Hormonal control of peroxidase activity in cultured *Pelargonium* pith. *Am. J. Bot.* **55,** 890–3. *86*

Lavender, D. P., Sweet, G. B., Zaerr, J. B. & Hermann, R. K. (1973). Spring shoot growth in Douglas-fir may be initiated by gibberellins exported from the roots. *Science* **182,** 838–9. *8, 61*

Leather, G. R., Forrence, L. E. & Abeles, F. B. (1972). Increased ethylene production during clinostat experiments may cause leaf epinasty. *Plant Physiol.* **49,** 183–6. *32, 101*

Lee, T. T. (1972). Interaction of cytokinin, auxin, and gibberellin on peroxidase isoenzymes in tobacco tissues cultured *in vitro*. *Can. J. Bot.* **50,** 2471–7. *86*

Leopold, A. C. (1972). Ethylene as a plant hormone. In: *Hormonal regulation in plant growth and development,* ed. H. Kaldewey & Y. Vardar, pp. 245–62. Weinheim: Verlag Chemie. *32, 101*

Lepp, N. W. & Peel, A. J. (1971*a*). Influence of IAA upon the longitudinal movement of labelled sugars in the phloem of willow. *Planta* **97,** 50–61. *61*

Lepp, N. W. & Peel, A. J. (1971*b*). Distribution of growth regulators and sugars by the tangential and radial transport systems of stem segments of willow. *Planta* **99,** 275–82. *61*

Leshem, B. (1966). Toxic effects of carbowaxes (polyethylene glycols) on *Pinus halepenis* Mill. seedlings. *Plant Soil* **24,** 322–4. *97*

Levitt, J. (1972). *Responses of plants to environmental stresses.* New York: Academic Press. *98*

Libbenga, K. R. & Torrey, J. G. (1973). Hormone-induced endoreduplication prior to mitosis in cultured pea root cortex cells. *Am. J. Bot.* **60,** 293–9. *49*

Linkins, A. E., Lewis, L. N. & Palmer, R. L. (1973). Hormonally induced changes in the stem and petiole anatomy and cellulase enzyme patterns in *Phaseolus vulgaris* L. *Plant Physiol.* **52,** 554–60. *34*

Lipetz, J. (1970). Wound-healing in higher plants. *Int. Rev. Cytol.* **27,** 1–28. *3*

Lipetz, J. & Garro, A. J. (1965). Ionic effects on lignification and peroxidase in tissue cultures. *J. Cell Biol.* **25,** 109–11. *88*

List, A., Jr. (1963). Some observations on DNA content and cell and nuclear volume growth in the developing xylem cells of certain higher plants. *Am. J. Bot.* **50,** 320–9. *49*

Livne, A. & Vaadia, Y. (1972). Water deficits and hormone relations. In: *Water deficits and plant growth,* vol. 3, ed. T. T. Kozlowski, pp. 255–75. New York: Academic Press. *98*

Lodewick, J. E. (1928). Seasonal activity of the cambium in some northeastern trees. *N.Y. State Coll. Forestry Syracuse Bull., Tech. Pub.* No. 23. *56*

Loomis, R. S. & Torrey, J. G. (1964). Chemical control of vascular cambium initiation in isolated radish roots. *Proc. Nat. Acad. Sci.* (USA) **52,** 3–11. *9, 32, 66, 67*

Lorz, A. P. (1947). Supernumerary chromonemal reproductions: polytene chromosomes, endomitosis, multiple chromosome complexes, polysomaty. *Bot. Rev.* **13,** 597–624. *49*

Lyne, R. L. & ap Rees, T. (1971). Invertase and sugar content during differentiation of roots of *Pisum sativum*. *Phytochem.* **10,** 2593–9. *90*

Lyon, C. J. (1972). Auxin control for orientation of pea roots grown on a clinostat or exposed to ethylene. *Plant Physiol.* **50,** 417–20. *32, 101*

McArthur, I. C. S. & Steeves, T. A. (1972). An experimental study of vascular differentiation in *Geum chiloense* balbis. *Bot. Gaz.* **133,** 276–87. *69*

Macchia, F. (1967). Effetti morfologici, anatomici ed ultrastrutturali indotti dai ritardanti di crescita su plantule di *Pisum sativum* L. var. 'Gloria di Quimper' coltivate in soluzione nutritizia. *Nuovo G. Bot. Ital.* **101,** 361–90. *107*

Macklon, A. E. S. & Weatherley, P. E. (1965). Controlled environment studies of the nature and origins of water deficits in plants. *New Phytol.* **64,** 414–27. *97*

McManus, M. A. & Roth, L. E. (1965). Fibrillar differentiation in myxomycete plasmodia. *J. Cell Biol.* **25,** 305–18. *71*

Maheshwari, M. G. & Noodén, L. D. (1971). A requirement for DNA synthesis during auxin induction of cell enlargement in tobacco pith tissue. *Physiol. Plant.* **24,** 282–7. *47*

Mahlberg, P. G., Olson, K. & Walkinshaw, C. (1970). Development of peripheral vacuoles in plant cells. *Am. J. Bot.* **57,** 962–86. *78*

Mahlberg, P. G., Olson, K. & Walkinshaw, C. (1971). Origin and development of plasma membrane derived invaginations in *Vinca rosea* L. *Am. J. Bot.* **58,** 407–16. *78*

Mahmood, A. (1968). Cell grouping and primary wall generations in the cambial zone, xylem, and phloem in *Pinus. Aust. J. Bot.* **16,** 177–95. *54*

Maitra, S. C. & De, D. N. (1971). Role of microtubules in secondary thickening of differentiating xylem element. *J. Ultrastruct. Res.* **34,** 15–22. *70, 71, 74, 79*

Mäkelä, O. & Nossal, G. J. V. (1962). Autoradiographic studies on the immune response. II. DNA synthesis amongst single antibody-producing cells. *J. Exp. Med.* **115,** 231–44. *46*

Maksymowych, R. (1973). *Analysis of leaf development.* London: Cambridge University Press. *72*

Maksymowych, R. & Kettrick, M. A. (1970). DNA synthesis, cell division, and cell differentiation during leaf development of *Xanthium pennsylvanicum. Am. J. Bot.* **57,** 844–9. *47*

Malamud, D. (1971). Differentiation and the cell cycle. In: *The cell cycle and cancer,* ed. R. Baserga, pp. 132–41. New York: Marcel Dekker. *46*

Marchant, R. & Robards, A. W. (1968). Membrane systems associated with the plasmalemma of plant cells. *Ann. Bot.* **32,** 457–70. *78*

Maretzki, A. & Nickell, L. G. (1973). Formation of protoplasts from sugarcane cell suspensions and the regeneration of cell cultures from protoplasts. In: *Protoplastes et fusion de cellules somatiques végétales.* Colloques Internationaux CNRS no. 212, pp. 51–63. Paris:CNRS. *115*

Marx-Figini, M. (1971). Investigations on biosynthesis of cellulose: $\overline{DP_w}$ and yield of cellulose of the alga *Valonia* in the presence of colchicine. *Biochim. Biophys. Acta* **237,** 76–7. *71*

Matile, Ph. & Winkenbach, F. (1971). Function of lysosomes and lysosomal enzymes in the senescing corolla of the morning glory (*Ipomoea purpurea*). *J. Exp. Bot.* **22,** 759–71. *79*

Matthysse, A. G. (1970). Organ specificity of hormone-receptor-chromatin interactions. *Biochim. Biophys. Acta* **199,** 519–21. *12, 40*

Matthysse, A. G. & Abrams, M. (1970). A factor mediating interaction of kinins with the genetic material. *Biochim. Biophys. Acta* **199,** 511–18. *12, 40*

Matthysse, A. G. & Phillips, C. (1969). A protein intermediary in the interaction of a hormone with the genome. *Proc. Nat. Acad. Sci.* (USA) **63,** 897–903. *12*

Matthysse, A. G. & Torrey, J. G. (1967). Nutritional requirements for polyploid mitoses in cultured pea root segments. *Physiol. Plant.* **20,** 661–72. *24, 48*

Maybrook, A. C. (1917). On the haustoria of *Pedicularis vulgaris. Ann. Bot.* **31,** 499–511. *111*

Mazia, D. (1967). Fibrillar structure in the mitotic apparatus. *Symp. Int. Soc. Cell Biol.* **6,** 39–54. *110*

Meins, F., Jr. (1974). Cell division and determination phase of cytodifferentiation in plants. In: *Results and problems in cell differentiation,* ed. W. Berrmann, J. Reinert & H. Ursprung. New York: Springer-Verlag (in press). *50*

Mia, A. J. (1970). Fine structure of active, dormant, and aging cambial cells in *Tilia americana*. *Wood Sci.* **3,** 34–42. *78*

Michel, B. E. (1966). Carbowax toxicity. *Proc. Assoc. Southern Agric. Workers* **63,** 293–4. *97*

Miller, C. O. (1961). A kinetin-like compound in maize. *Proc. Nat. Acad. Sci.* (USA) **47,** 170–4. *35, 36*

Miller, J. H. & Stephani, M. C. (1971). Effects of colchicine and light on cell form in fern gametophytes. Implications for a mechanism of light-induced cell elongation. *Physiol. Plant.* **24,** 264–71. *101*

Minocha, S. C. & Halperin, W. (1973). Enzymatic changes during xylogenesis of cultured Jerusalem artichoke tissue. *Plant Physiol.* **51** (suppl.), 22. *86*

Minocha, S. C. & Halperin, W. (1974). Hormones and metabolites which control tracheid differentiation, with or without concomitant effects on growth, in cultured tuber tissue of *Helianthus tuberosus* L. *Planta* **116,** 319–31. *25, 36, 92, 93*

Mitchison, J. M. (1969). Markers in the cell cycle. In: *The cell cycle. Gene-enzyme interactions,* ed. G. M. Padilla, G. L. Whitson & I. L. Cameron, pp. 361–72. New York: Academic Press. *44*

Mitchison, J. M. (1971). *The biology of the cell cycle.* London: Cambridge University Press. *40, 44, 47*

Mitchison, J. M. (1973). Differentiation in the cell cycle. In: *The cell cycle in development and differentiation,* ed. M. Balls & F. S. Billett, pp. 1–11. London: Cambridge University Press. *44, 45*

Mitchison, J. M. & Creanor, J. (1969). Linear synthesis of sucrase and phosphatases during the cell cycle of *Schizosaccharomyces pombe*. *J. Cell Sci.* **5,** 373–91. *41*

Mizuno, K., Komamine, A. & Shimokoriyama, M. (1971). Vessel element formation in cultured carrot-root phloem slices. *Plant Cell Physiol.* **12,** 823–30. *100*

Morey, P. R. (1973). The effects of DPX-1840 on cambial cell differentiation. *Am. J. Bot.* **60** (suppl.), 11. *9, 64, 108*

Morey, P. R. (1974). Influence of 3,3a-dihydro-2-(*p*-methoxyphenyl)-8*H*-pyrazolo-(5,1-a)isoindol-8-one on xylem formation in honey mesquite. *Weed Sci.* **22,** 6–10. *9*

Murashige, T. (1964). Analysis of the inhibition of organ formation in tobacco tissue culture by gibberellin. *Physiol. Plant.* **17,** 636–43. *28*

Murashige, T. & Skoog, F. (1962). A revised medium for rapid growth and bioassays with tobacco tissue cultures. *Physiol. Plant.* **15.** 473–97. *17, 19, 21, 27, 28, 100*

Murmanis, L. (1970). Locating the initial in the vascular cambium of *Pinus strobus* L. by electron microscopy. *Wood Sci. Tech.* **4,** 1–14. *78*

Murmanis, L. (1971*a*). Structural changes in the vascular cambium of *Pinus strobus* L. during an annual cycle. *Ann. Bot.* **35,** 133–41. *78*

Murmanis, L. (1971*b*). Particles and microtubules in vascular cells of *Pinus strobus* L. during cell wall formation. *New Phytol.* **70,** 1089–93. *71, 78*

Murmanis, L. & Sachs, I. B. (1969). Seasonal development of secondary xylem in *Pinus strobus* L. *Wood Sci. Tech.* **3,** 177–93. *78*

Murmanis, L. & Sachs, I. B. (1973). Cell wall formation in secondary xylem of *Pinus strobus* L. *Wood Sci. Tech.* **7,** 173–88. *78, 79*

Nagl, W. (1968*a*). Der mitotische und endomitotische Kernzyklus bei *Allium carinatum*. I. Struktur, Volumen und DNS-Gehalt der Kerne. *Öst. Bot. Z.* **115,** 322–53. *40*

Nagl, W. (1968*b*). Die Kernstruktur während des mitotischen und endomitotischen Zellzyklus *Ber. Dtsch. Bot. Ges.* **81,** 320–4. *40*

Nagl, W. (1970). Correlation of chromatin structure and interphase stage in nuclei of *Allium flavum*. *Cytobiol.* **1,** 395–8. *40*

Nagl, W. (1972). Selective inhibition of cell cycle stages in the *Allium* root meristem by colchicine and growth regulators. *Am. J. Bot.* **59,** 346–51. *49*

Nagl, W. (1973). The mitotic and endomitotic nuclear cycle in *Allium carinatum*. IV. ^{3}H-uridine incorporation. *Chromosoma* **44,** 203–12. *50*

Nagl, W., Hendon, J. & Rücker, W. (1972). DNA amplification in *Cymbidium* protocorms *in vitro,* as it relates to cytodifferentiation and hormone treatment. *Cell Differ.* **1,** 229–37. *51, 52*

Nagl, W. & Rücker, W. (1972). Beziehungen zwischen Morphogenese und nuklearem DNS-Gehalt bei aseptischen Kulturen von *Cymbidium* nach Wuchsstoffbehandlung. *Z. Pflanzenphysiol.* **67,** 120–34. *49, 50, 52*

Naik, G. G. (1965). Studies on the effects of temperature on the growth of plant tissue cultures. MS Thesis, University of Edinburgh, Scotland. *100*

Naylor, A. W. (1972). Water deficits and nitrogen metabolism. In: *Water deficits and plant growth,* vol. 3, ed. T. T. Kozlowski, pp. 241–54. New York: Academic Press. *98*

Neel, P. L. (1970). Experimental manpulation of trunk growth in young trees. *Proc. 46th Int. Shade Tree Conference,* ed. D. Neely, pp. 25–36. Urbana: International Shade Tree Conference Inc. *32*

Neish, A. C. (1964). Cinnamic acid derivatives as intermediates in the biosynthesis of lignin and related compounds. In: *The formation of wood in forest trees,* ed. M. H. Zimmermann, pp. 219–39. New York: Academic Press. *81, 82*

Neish, A. C. (1965). Coumarins, phenylpropanes, and lignin. In: *Plant biochemistry,* ed. J. Bonner & J. E. Varner, pp. 581–617. New York: Academic Press. *82*

Nelmes, B. J., Preston, R. D. & Ashworth, D. (1973). A possible function of microtubules suggested by their abnormal distribution in rubbery wood. *J. Cell Sci.* **13,** 741–51. *71*

Newcomb, E. H. (1969). Plant microtubules. *Annu. Rev. Plant Physiol.* **20,** 253–88. *71*

Niedergang-Kamien, E. & Leopold, A. C. (1957). Inhibitors of polar auxin transport. *Physiol. Plant.* **10,** 29–38. *107*

Nitsch, J. P. & Nitsch, C. (1960). Le problème de l'action des auxines sur la division cellulaire: présence d'un cofacteur de division dans le tubercule de Topinambour. *Ann. Physiol. Vég.* **2,** 261–8. *21, 27*

Noodén, L. D. (1971). Physiological and developmental effects of colchicine. *Plant Cell Physiol.* **12,** 759–70. *72*

Northcote, D. H. (1968). The organization of the endoplasmic reticulum, the Golgi bodies and microtubules during cell division and subsequent growth. In: *Plant cell organelles,* ed. J. B. Pridham, pp. 179–97. New York: Academic Press. *93, 94*

Northcote, D. H. (1969*a*). The synthesis and metabolic control of polysaccharides and lignin during the differentiation of plant cells. In: *Essays in biochemistry,* vol. 5, ed. P. N. Campbell & G. D. Greville, pp. 89–137. New York: Academic Press. *93, 94*

Northcote, D. H. (1969*b*). Fine structure of cytoplasm in relation to synthesis and secretion in plant cells. *Proc. Royal Soc., Ser. B.* **173,** 21–30. *71, 76, 94*

Northcote, D. H. (1971). Organization of structure, synthesis, and transport within the plant during cell division and growth. In: *Control mechanisms of growth and differentiation,* ed. D. D. Davies and M. J. Balls, pp. 51–69. London: Cambridge University Press. *76, 93, 94*

Nuti Ronchi, V. (1971). Amplificazione genica in cellule sdifferenziate di cotiledoni di *Lactuca sativa* coltivati *in vitro. Atti Assoc. Genet. Ital.* **16,** 47–50. *51*

O'Brien, T. P. (1970). Further observations on hydrolysis of the cell wall in the xylem. *Protoplasma* **69,** 1–14. *76, 77*

O'Brien, T. P. (1972). The cytology of the cell-wall formation in some eukaryotic cells. *Bot. Rev.* **38,** 87–118. *74, 76, 82*

O'Brien, T. P. (1974). Primary vascular tissues. In: *Dynamic aspects of plant ultrastructure,* ed. A. W. Robards, pp. 414–40. New York: McGraw-Hill. *70, 74, 77, 82*

O'Brien, T. P. & McCully, M. E. (1969). *Plant structure and development. A pictorial and physiological approach.* New York: Macmillan. *75*

O'Brien, T. P. & Thimann, K. V. (1967). Observations on the fine structure of the oat coleoptile. III. Correlated light and electron microscopy of the vascular tissues. *Protoplasma* **63,** 443–78. *77*

Obroucheva, N. V. (1969). The relationship between growth and lignification in young maize roots. *Doklady Acad. Nauk USSR* **185,** 1178–81. *84*

Osborne, D. J., Ridge, I. & Sargent, J. A. (1972). Ethylene and the growth of plant cells: role of peroxidase and hydroxyproline-rich proteins. In: *Plant growth substances 1970,* ed. D. J. Carr, pp. 534–42. New York: Springer-Verlag. *34, 86*

Parish, R. W. (1968). Studies on senescing tobacco leaf disks with special reference to peroxidase. II. The effects and interactions of proline, hydroxyproline and kinetin. *Planta* **82,** 14–21. *86*

Partanen, C. R. (1965). On the chromosomal basis for cellular differentiation. *Am. J. Bot.* **52,** 204–9. *49*

Parups, E. V. (1973). Control of ethylene-induced responses in plants by a substituted benzothiodiazole. *Physiol. Plant.* **29,** 365–70. *106*

Pate, J. S., Gunning, B. E.S. & Milliken, F. F. (1970). Function of transfer cells in the nodal region of stems, particularly in relation to the nutrition of young seedlings. *Protoplasma* **71,** 313–34. *112*

Patil, S. S. & Tang, C. S. (1974). Inhibition of ethylene evolution in papaya pulp tissue by benzyl isothiocyanate. *Plant Physiol.* **53,** 585–8. *35, 106*

Pearson, M. J. (1974). Polyteny and the functional significance of the polytene cell cycle. *J. Cell Sci.* **15,** 457–79. *49*

Péaud-Lenoël, C. & Jouanneau, J. P. (1971). Contrôle du cycle mitotique dans les suspensions de cellules de tabac cultivées en milieu liquide. In: *Les cultures de tissus de plantes.* Colloques Internationaux CNRS no. 193, pp. 95–102. Paris:CNRS. *48*

Peel, A. J. (1964). Tangential movement of ^{14}C-labelled assimilates in stems of willow. *J. Exp. Bot.* **15,** 104–13. *61*

Peel, A. J. (1966). The sugars concerned in the tangential movement of ^{14}C-labelled assimilates in willow. *J. Exp. Bot.* **17,** 156–64. *61*

Peer, H. G. (1971). Degradation of sugars and their reactions with amino acids. In: *Effects of sterilization on components in nutrient media,* ed. J. Van Bragt *et al.*, pp. 105–15. Wageningen: H. Veenman & Zonen N. V. *91*

Peterson, R. L. (1973). Control of cambial activity in roots of turnip (*Brassica rapa*). *Can. J. Bot.* **51,** 475–80. *9, 66, 68*

Philipson, W. R., Ward, J. M. & Butterfield, B. G. (1971). *The vascular cambium. Its development and activity.* London: Chapman & Hall. *54, 55, 111*

Phillips, R. & Torrey, J. G. (1973). DNA synthesis, cell division and specific cytodifferentiation in cultured pea root cortical explants. *Develop. Biol.* **31,** 336–47 *2, 7, 20, 42–3, 49, 50, 115*

Phillips, R. & Torrey, J. G. (1974). DNA levels in differentiating tracheary elements. *Develop. Biol.* **39,** 322–5. *42, 50, 51, 52*

Pickett-Heaps, J. D. (1966). Incorporation of radioactivity into wheat xylem walls. *Planta,* **1,** 1–14. *76*

Pickett-Heaps, J. D. (1967). The effects of colchicine on the ultrastructure of dividing plant cells, xylem wall differentiation and distribution of cytoplasmic microtubules. *Develop. Biol.* **15,** 206–36. *71, 76*

Pickett-Heaps, J. D. (1968). Xylem wall deposition. Radioautographic investigations using lignin precursors. *Protoplasma* **65,** 181–205. *71, 74, 82*

Pickett-Heaps, J. D. (1974). Plant microtubules. In: *Dynamic aspects of plant ultrastructure,* ed. A. W. Robards, pp. 219–55. New York: McGraw-Hill. *71, 74*

Pickett-Heaps, J. D. & Northcote, D. H. (1966*a*). The relationship of cellular organelles to the formation and development of the plant cell wall. *J. Exp. Bot.* **17,** 20–6. *71, 76*

Pickett-Heaps, J. D. & Northcote, D. H. (1966*b*). Cell division in the formation of the stomatal complex of the young leaves of wheat. *J. Cell Sci.* **1,** 121–8. *38*

Pieniazek, J. & Saniewski, M. (1968). The synergistic effect of benzyladenine and morphactin on cambial activity in apple shoots. *Bull. Acad. Pol. Sci.* **16,** 381–4. *109*

Pieniazek, J., Smolinski, M. & Saniewski, M. (1970). Induced structural changes in anatomy of apple shoots after treatment with morphactin IT-3456 and other growth regulators (NAA, GA, BA). *Acta Agrobot.* **23,** 387–96. *109*

Pierik, R. L. M. (1971). Plant tissue culture as motivation for the symposium. In: *Effects of sterilization on components in nutrient media,* ed. J. van Bragt, D. A. A. Mossel, R. L. M. Pierik & H. Veldstra, pp. 3–13. Wageningen: H. Veenman & Zonen N. V. *89*

Priestley, J. H., Scott, L. I. & Malins, M. E. (1933). A new method of studying cambial activity. *Leeds Phil. Lit. Soc., Sci. Sect. Proc.* **2,** 365–74. *56*

Priestley, J. H., Scott, L. I. & Malins, M. E. (1935). Vessel development in the angiosperms. *Leeds Phil. Lit. Soc., Sci. Sect. Proc.* **3,** 42–54. *56*

Quastler, H. (1963). The analysis of cell population kinetics. In: *Cell proliferation,* ed. L. F. Lamberton & R. J. Fry, pp. 18–34. Philadelphia: F. A. Davis Co. *39*

Radin, J. W. & Loomis, R. S. (1969). Ethylene and carbon dioxide in the growth and development of cultured radish roots. *Plant Physiol.* **44,** 1584–9. *32, 106*

Radin, J. W. & Loomis, R. S. (1971). Changes in the cytokinins of radish roots during maturation. *Physiol. Plant.* **25,** 240–4. *22, 66*

Rappaport, L. & Sachs, M. (1967). Wound-induced gibberellin. *Nature,* (Lond.) **214,** 1149–50. *28*

Ray, P. M. (1973*a*). Regulation of β-glucan synthetase activity by auxin in pea stem tissue. I. Kinetic aspects. *Plant Physiol.* **51,** 601–8. *16*

Ray, P. M. (1973*b*). Regulation of β-glucan synthetase activity by auxin in pea stem tissue. II. Metabolic requirements. *Plant Physiol.* **51,** 609–14. *16*

Ray, P. M., Shininger, T. L. & Ray, M. M. (1969). Isolation of β-glucan synthetase particles from plant cells and identification with Golgi membranes. *Proc. Nat. Acad. Sci.* (USA) **64,** 605–12. *74*

Rayle, D. L. & Cleland, R. (1970). Enhancement of wall loosening and elongation by acid solutions. *Plant Physiol.* **46,** 250–3. *103*

Rédei, G. P. (1974). 'Fructose effect' in higher plants. *Ann. Bot.,* **38,** 287–97. *91*

Reinders-Gouwentak, C. A. (1941). Cambial activity as dependent on the presence of growth hormone and the non-resting condition of stems. *Proc. Ned. Akad. Wetensch. Amst.* **44,** 654–62. *8, 62*

Reines, M. (1959). The initiation of cambial activity in black cherry. *Forest Sci.* **5,** 70–3. *57*

Reinhold, L., Sachs, T. & Vislovska, L. (1972). The role of auxin in thigmotropism. In: *Plant growth substances, 1970,* ed. D. J. Carr, pp. 731–7. New York: Springer-Verlag. *102*

Ricardo, C. P. P. & ap Rees, T. (1970). Invertase activity during the development of carrot roots. *Phytochem.* **9,** 239–47. *90*

Ridge, I. (1973). The control of cell shape and rate of cell expansion by ethylene: effects on microfibril orientation and cell wall extensibility in etiolated peas. *Acta Bot. Neerl.* **22,** 144–58. *86*

Ridge, I. & Osborne, D. J. (1970). Hydroxyproline and peroxidases in cell walls of *Pisum sativum:* regulation by ethylene. *J. Exp. Bot.* **21,** 843–56. *32, 86*

Ridge, I & Osborne, D. J. (1971). Role of peroxidase when hydroxyproline-rich protein in plant cell walls is increased by ethylene. *Nature New Biol.* **229,** 205–8. *32*

Rier, J. P. (1970). Chemical basis for vascular tissue differentiation in plant tissues. *Trans. N. Y. Acad. Sci.* **32,** 594–9. *89*

Rier, J. P. & Beslow, D. T. (1967). Sucrose concentration and the differentiation of xylem in callus. *Bot. Gaz.* **128,** 73–7. *7*

Rier, J. P. & Owens, V. A. (1973). Ozone and xylem regeneration in internodes of *Coleus. Am. J. Bot.* **60**(suppl.), 12. *103*

Ritzert, R. W. & Turin, B. A. (1970). Formation of peroxidases in response to indole-3-acetic acid in cultured tobacco cells. *Phytochem.* **9.** 1701–5. *86*

Robards, A. W. (1968). On the ultrastructure of differentiating secondary xylem in willow. *Protoplasma* **65,** 449–64. *71, 78*

Robards, A. W. (1970). *Electron microscopy and plant ultrastructure.* New York: McGraw-Hill. *70, 76*

Robards, A. W., Davidson, E. & Kidwai, P. (1969). Short-term effects of some chemicals on cambial activity. *J. Exp. Bot.* **20,** 912–20. *61*

Robards, A. W. & Kidwai, P. (1969*a*). Cytochemical localization of phosphatase in differentiating secondary vascular cells. *Planta* **87,** 227–38. *86*

Robards, A. W. & Kidwai, P. (1969*b*). A comparative study of the ultrastructure of resting and active cambium of *Salix fragilis* L. *Planta* **84,** 239–49. *78*

Robards, A. W. & Kidwai, P. (1972). Microtubules and microfibrils in xylem fibres during secondary cell wall formation. *Cytobiol.* **6,** 1–21. *72, 78, 79*

Roberts, L. W. (1969). The initiation of xylem differentiation. *Bot. Rev.* **35,** 201–50. *3, 8, 70, 82, 88, 107, 115*

Roberts, L. W. & Baba, S. (1968*a*). IAA-induced xylem differentiation in the presence of colchicine. *Plant Cell Physiol.* **9,** 315–21. *71*

Roberts, L. W. & Baba, S. (1968*b*). Effect of proline on wound vessel member formation. *Plant Cell Physiol.* **9,** 353–60. *34, 86*

Roberts, L. W. & Baba, S. (1970). Auxin and kinetin interaction during xylem differentiation. *Mem. Fac. Science, Kyoto Univ., Ser. Biol.* **3,** 1–12. *14, 15*

Roberts, L. W. & Fosket, D. E. (1962). Geotropic stimulation: effects on wound vessel differentiation. *Science* **138,** 1264–5. *32, 101*

Roberts, L. W. & Fosket, D. E. (1966). Interaction of gibberellic acid and indoleacetic acid in the differentiation of wound vessel members. *New Phytol.* **65,** 5–8. *7*

Roberts, L. W. & Sankhla, N. (1973). Inhibition of xylogenesis by morphactin in pith parenchyma explants of *Lactuca. Plant Cell Physiol.* **14,** 521–30. *107, 109, 110*

Robnett, W. E. & Morey, P. R. (1973). Wood formation in *Prosopis:* effect of 2,4-D, 2,4,5-T, and TIBA. *Am. J. Bot.* **60,** 745–54. *108*

Romberger, J. A. & Tabor, C. A. (1971). The *Picea abies* shoot apical meristem in culture. I. Agar and autoclaving effects. *Am. J. Bot.* **58,** 131–40. *90.*

Rossini, L. (1973). Sur les différences d'action de l'acide 2, 4-dichlorophenoxyacétique et de la 6-benzyladenine sur la division *in vitro* des cellules du parenchyme foliaire de *Calystegia sepium* (L.) R. Br. *C.R. Acad. Sci.* (Paris) **276,** 1689–92. *25*

Roy-Burnman, P. (1970). *Analogues of nucleic acid components: mechanisms of action. Recent results in cancer research,* vol. 25. New York: Springer-Verlag. *106*

Rubery, P. H. (1972). The activity of uridine diphosphate-D-glucose: nicotinamide-adenine dinucleotide oxidoreductase in cambial tissue and differentiating xylem isolated from sycamore trees. *Planta* **103,** 188–92. *93*

Rubery, P. H. (1973). The activity of uridine diphosphate-D-glucose-4-epimerase in cambial tissue and differentiating xylem isolated from sycamore (*Acer pseudoplatanus*) trees. *Planta* **111,** 267–9. *93*

Rubery, P. H. & Fosket, D. E. (1969). Changes in phenylalanine ammmonia-lyase during xylem differentiation in *Coleus* and soybean. *Planta* **87,** 54–62. *84*

Ruf, R. H., Eckert, R. E. & Gifford, R. O. (1963). Osmotic adjustment of cell sap to increases in root medium osmotic stress. *Soil Sci.* **96,** 326–30. *97*

Runner, M. N. (1970). Changing syntheses in development. *Changing syntheses in development,* ed. M. N. Runner, pp. 1–11. New York: Academic Press. *45*

Rytömaa, T. (1973). Control of cell division in mammalian cells. In: *The Cell cycle in development and differentiation,* ed. M. Balls & F. S. Billett, pp. 457–72. London: Cambridge University Press. *44*

Sabnis, D. D., Hirshberg, G. & Jacobs, W. P. (1969). Radioautographic analysis of the distribution of label from ^{3}H-indoleacetic acid supplied to isolated *Colus* internodes. *Plant Physiol.* **44,** 27–36. *85*

Sachs, T. (1968). On the determination of the pattern of vascular tissues in peas. *Ann. Bot.* **32,** 781–90. *16*

Sachs, T. (1969). Polarity and the induction of organized vascular tissues. *Ann. Bot.* **33,** 263–75. *16, 17, 45*

Safwat, F. M. (1969). The initiation of vascular cambium and production of secondary xylem in flower bud pedicels of *Asclepias curassavica* L. in culture. *Ann. Mo. Bot. Gard.* **56,** 251–60. *68*

Sampson, M. & Davies, D. D. (1966). Synthesis of a metabolically labile DNA in the maturing root cells of *Vicia faba*. *Exp. Cell Res.* **43,** 669–73. *52*

Saniewski, M., Smolinski, M. & Pieniazek, J. (1968). The effect of morphactin on the anatomical structures of *Pisum sativum* L. and *Dolichos lablab* L. roots. *Bull. Acad. Pol. Sci.* **16,** 513–15. *109*

Sanio, K. (1873). Anatomie der gemeinen Kiefer (*Pinus silvestris* L.). *Jahrb. wiss. Bot.* **9,** 50–126. *54*

Sargent, J. A., Atack, A. V. & Osborne, D. J. (1973). Orientation of cell growth in the etiolated pea stem. Effect of ethylene and auxin on cell wall deposition. *Planta* **109,** 185–92. *34, 86*

Sargent, J. A., Atack, A.V. & Osborne, D. J. (1974). Auxin and ethylene control of growth in epidermal cells of *Pisum sativum:* a biphasic response to auxin. *Planta* **115,** 213–25. *34*

Sarkanen, K. V. & Ludwig, C. H. (eds) (1971). *Lignins. Occurrence, formation, structrue, and reactions.* New York: Wiley-Interscience. *82*

Saussay, R. (1969). Action de l'acide gibbérellique sur les phénomènes d'histogenèse dans les tissus du cambium de Saule (*Salix cinerea* L.) cultivés *in vitro. C. R. Acad. Sci.* (Paris) **269,** 167–70. *28, 64*

Sauter, J. J. (1972). Cytochemical demonstration of sulfhydryl disulfide-containing proteins in sieve elements of conifers. *Naturwiss.* **59,** 470. *85*

Sauter, J. J., Iten, W. & Zimmermann, M. H. (1973). Studies on the release of sugar into the vessels of sugar maple (*Acer saccharum*). *Can. J. Bot.* **51,** 1–8. *85*

Sawhney, V. K. & Srivastava, L. M. (1974). Gibberellic acid induced elongation of lettuce hypocotyls and its inhibition by colchicine. *Can. J. Bot.* **52,** 259–64. *29*

Schneider, G. (1970). Morphactins: physiology and performance. *Annu. Rev. Plant Physiol.* **21,** 499–536. *107, 109*

Schrank, A. R. (1959). Electronasty and electrotropism. In: *Encyclopedia of plant physiology,* vol. 17(1), ed. W. Ruhland, pp. 148–63. Berlin: Springer-Verlag. *102*

Schröter, K. & Sievers, A. (1971). Effect of turgor reduction on the Golgi apparatus and the cell wall formation in root hairs. *Protoplasma* **72,** 203–11. *96*

Schubert, W. J. (1965). *Lignin biochemistry*. New York: Academic Press. *82*

Scurfield, G. & Bland, D. E. (1963). The anatomy and chemistry of 'rubbery wood' in apple var. Lord Lambourne. *J. Hort. Sci.* **38,** 297–306. *71*

Sexton, R. & Sutcliffe, J. F. (1969). The distribution of β-glycerophosphatase in young roots of *Pisum sativum* L. *Ann. Bot.* **33,** 407–19. *85*

Shain, L. & Hillis, W. E. (1973). Ethylene production in xylem of *Pinus radiata* in relation to heartwood formation. *Can. J. Bot.* **51,** 1331–5. *32*

Shaykh, M. M. & Roberts, L. W. (1974). A histochemical study of phosphatases in root apical meristems. *Ann. Bot.* **38,** 165–74. *85*

Sheldrake, A. R. (1970). Cellulase and cell differentiation in *Acer pseudoplatanus*. *Planta* **95,** 167–78. *77*

Sheldrake, A. R. (1971). Auxin in the cambium and its differentiating derivatives. *J. Exp. Bot.* **22,** 735–40. *58*

Sheldrake, A. R. (1973*a*). The production of hormones in higher plants. *Biol. Rev.* **48,** 509–59. *9, 31, 60, 61*

Sheldrake, A. R. (1973*b*). Auxin transport in secondary tissues. *J. Exp. Bot.* **24,** 87–96. *16*

Sheldrake, A. R. & Northcote, D. H. (1968*a*). Some constituents of xylem sap and their possible relationship to xylem differentiation. *J. Exp. Bot.* **19,** 681–9. *15, 77*

Sheldrake, A. R. & Northcote, D. H. (1968*b*). The production of auxin by tobacco internode tissues. *New Phytol.* **67,** 1–13. *15*

Shepherd, K. R. & Rowan, K. S. (1967). Indoleacetic acid in cambial tissue in radiata pine. *Aust. J. Biol. Sci.* **20,** 637–46. *59*

Shibaoka, H. (1972). Gibberellin-colchicine interaction in elongation of azuki epicotyl sections. *Plant Cell Physiol.* **13,** 461–9. *29*

Shibaoka, H. (1974). Involvement of wall microtubules in gibberellin promotion and kinetin inhibition of stem elongation. *Plant Cell Physiol.* **15,** 255–63. *25, 29*

Shininger, T. L. (1970). The production and differentiation of secondary xylem in *Xanthium pennsylvanicum*. *Am. J. Bot.* **57,** 769–81. *9, 54, 63, 64*

Shininger, T. L. (1971). The regulation of cambial division and secondary xylem differentiation in *Xanthium* by auxins and gibberellins. *Plant Physiol.* **47,** 417–22. *54, 108*

Shininger, T. L. & Torrey, J. G. (1974). The roles of cytokinins in the induction of cell division and cytodifferentiation in pea root cortical tissue *in vitro*. In: *Mechanisms of Regulation of Plant Growth,* ed. R. L. Bieleski, A. R. Ferguson & M. M. Cresswell, pp. 721–8. Wellington: Royal Society of New Zealand. *22, 24*

Siebers, A. M. (1971*a*). Initiation of radial polarity in the interfascicular cambium of *Ricinus communis* L. *Acta Bot. Neerl.* **20,** 211–20. *2, 11, 45, 64, 68*

Siebers, A. M. (1971*b*). Differentiation of isolated interfascicular tissue of *Ricinus communis* L. *Acta Bot. Neerl.* **20,** 343–55. *7, 9, 45, 64, 68–9*

Siebers, A. M. (1972). Vascular bundle differentiation and cambial development in cultured tissue blocks excised from the embryo of *Ricinus communis* L. *Acta Bot. Neerl.* **21,** 327–42. *68, 69*

Siebers, A. M. & Ladage, C. A. (1973). Factors controlling cambial development in the hypocotyl of *Ricinus communis* L. *Acta Bot. Neerl.* **22,** 416–32. *28, 107, 108*

Simon, S. (1908*a*). Experimentelle Untersuchungen über die Differenzierungsvorgänge im Callusgewebe von Holzgewachsen. *Jarhb. wiss. Bot.* **45,** 351–478. *4*

Simon, S. (1908*b*). Experimentelle Untersuchungen über die Entstehung von Gefässvergindungen. *Ber. Dtsch. Bot. Ges.* **26,** 364–96. *7*

Simpson, P. G. & Fineran, B. A. (1970). Structure and development of the haustorium in *Mida salicifolia*. *Phytomorphology* **20,** 236–48. *111*

Sinnott, E. W. & Bloch, R. (1944). Visible expression of cytoplasmic pattern in the differentiation of xylem strands. *Proc. Nat. Acad. Sci.* (USA) **30,** 388–92. *4, 71*

Sinnott, E. W. & Bloch, R. (1945). The cytoplasmic basis of intercellular patterns in vascular differentiation. *Am. J. Bot.* **32,** 151–6. *4, 71*

Skene, D. S. (1969). The period of time taken by cambial derivatives to grow and differentiate into tracheids in *Pinus radiata*. *Ann. Bot.* **33,** 253–62. *2, 63, 76*

Smith, J. A. & Martin, L. (1973). Do cells cycle? *Proc. Nat. Acad. Sci.* (USA) **70,** 1263–7. *39*

Smolinski, M., Saniewski, M. & Pieniazek, J. (1972). The effort of morphactin IT-3456 on cambial activity and wood differentiation in *Picea excelsa*. *Bull. Acad. Pol. Sci.* **20,** 431–5. *109*

Snijman, D. A. (1972). Tracheary element differentiation in *in vitro* grown *Nicotiana tabacum* L. callus. MS Thesis, University of Natal, Pietermaritzburg. *28*

Söding, H. (1937). Wuchsstoff und Kambiumtätigkeit. *Jahrb. wiss. Bot.* **84,** 639–70. *8, 59*

Soe, K. (1959). Morphogenetic studies on *Onoclea sensibilis* L. PhD Thesis, Harvard University, Cambridge, Mass. *69*

Sorokin, H. P., Mathur, S. N. & Thimann, K. V. (1962). The effects of auxins and kinetin on xylem differentiation in the pea epicotyl. *Am. J. Bot.* **49,** 444–53. *8*

Spencer, F. S., Ziola, B. & Maclachlan, G. A. (1971). Particulate glucan synthetase activity: dependence on acceptor, activator, and plant growth hormone. *Can. J. Biochem.* **49,** 1326–32. *16*

Srivastava, L. M. & O'Brien, T. P. (1966). On the ultrastructure of cambium and its vascular derivatives. I. Cambium of *Pinus strobus* L. *Protoplasma* **61,** 257–76. *77, 78*

Srivastava, L. M. & Singh, A. P. (1972). Certain aspects of xylem differentiation in corn. *Can. J. Bot.* **50,** 1795–1804. *70, 71, 77, 79*

Stafford, H. A. (1962). Histochemical and biochemical differences between lignin-like materials in *Phleum pratense* L. *Plant Physiol.* **37,** 643–9. *88*

Stebbins, G. L. & Shah, S. S. (1960). Developmental studies of cell differentiation in the epidermis of monocotyledons. II. Cytological features of stomatal development in the Gramineae. *Develop. Biol.* **2,** 477–500. *38*

Steeves, T. A. & Sussex, I. M. (1972). *Patterns in plant development*. Englewood Cliffs: Prentice-Hall. *10, 55, 64*

Stehsel, M. L. & Caplin, S. M. (1969). Sugars: autoclaving vs. sterile filtration on the growth of carrot root tissue in culture. *Life Sci.* **8,** 1255–9. *90, 91*

Stein, O. L., Rowley, J. R. & Lockhart, J. A. (1971). Deformation of cell shape and pit pattern in roots of *Zea mays* under the influence of colchicine and heavy water. *Phytomorphology* **21,** 296–308. *72*

Stetler, D. A. & DeMaggio, A. E. (1972). An ultrastructural study of fern gametophytes during one- to two-dimensional development. *Am. J. Bot.* **59,** 1011–17. *101*

Steward, F. C., Mapes, M. O. & Mears, K. (1958). Growth and organized development of cultured cells. II. Organization in cultures grown from freely suspended cells. *Am. J. Bot.* **45,** 705–8. *97*

Stewart, C. M. (1969). The formation and chemical composition of hardwoods. *Appita* **22,** 32–60. *82*

Stewart, C. M., Melvin, J. F., Ditchburne, N., Tham, S. H. & Zerdoner, E. (1973). The effect of season of growth on the chemical composition of cambial saps of *Eucalyptus regnans* trees. *Planta* **12,** 349–72. *96*

Stewart, G. R. & Smith, H. (1972). Effects of absisic acid on nucleic acid synthesis and the induction of nitrate reductase in *Lemna polyrhiza*. *J. Exp. Bot.* **23,** 875–85. *36*

Stockdale, F. E. & Holtzer, H. (1961). DNA synthesis and myogenesis. *Exp. Cell Res.* **24,** 508. *46*

Stockdale, F. E. & Topper, Y. J. (1966). The role of DNA synthesis and mitosis in hormone-dependent differentiation. *Proc. Nat. Acad. Sci.* (USA) **56,** 1283–9. *41*

Street, H. E. (ed.) (1974). *Plant tissue and cell culture.* Berkeley: University of California Press. *95*

Sussex, I. M. & Clutter, M. E. (1967). Differentiation in tissues, free cells, and reaggregated plant cells. *In Vitro* **3,** 3–12. *97*

Sussex, I. M., Clutter, M. E. & Goldsmith, M. H. M. (1972). Wound recovery by pith cell redifferentiation: structural changes. *Am. J. Bot.* **59,** 797–804. *16, 48*

Swain, L. W. & Rier, J. P. (1968). Tumorigenesis on mineral-deficient tomato plants. *Cancer Res.* **28,** 2496–2501. *89*

Swift, H. (1950). The constancy of deoxyribose nucleic acid in plant nuclei. *Proc. Nat. Acad. Sci.* (USA) **36,** 643–54. *50*

Syōno, K. & Furuya, T. (1971). Effects of temperature on the cytokinin requirement of tobacco calluses. *Plant Cell Physiol.* **12,** 61–71. *100*

Szalai, I. & Gracza, L. (1958). Quantitative distribution and changes of the free tryptophan in the twigs of the ash. *Phyton* **2,** 111–14. *56*

Tepper, H. B. & Hollis, C. A. (1967). Mitotic reactivation of the terminal bud and cambium of white ash. *Science* **156,** 1635–6. *59*

Thimann, K. V. (1972). Auxins. An informal summary of some recent work. In: *Hormonal regulation in plant growth and development,* ed. H. Kaldewey & Y. Vardar, pp. 155–70. Weinheim: Verlag Chemie. *55*

Thompson, N. P. (1967). The time course of sieve tube and xylem cell regeneration and their anatomical orientation in *Coleus* stem. *Am. J. Bot.* **54,** 588–95. *114*

Thompson, N. P. (1968). Polarity of IAA-^{14}C and 2,4-D-^{14}C transport and vascular regeneration in isolated internodes of peanut. In: *Biochemistry and physiology of plant growth substances,* ed. F. Wightman & G. Setterfield, pp. 1205–13. Ottawa: Runge Press. *18*

Thompson, N. P. (1970). The transport of auxin and regeneration of xylem in okra and pea stems. *Am. J. Bot.* **57,** 390–8. *17, 18*

Thompson, N.P. & Jacobs, W. P. (1966). Polarity of IAA effect on sieve-tube and xylem regeneration in *Coleus* and tomato stems. *Plant Physiol.* **41,** 673–82. *18*

Tilney, L. G. (1971). Origin and continuity of microtubules. In: *Origin and continuity of cell organelles,* ed. J. Reinert & H. Ursprung, pp. 222–60. New York: Springer-Verlag. *74*

Tilney, L. G. & Gibbins, J. R. (1969). Microtubules and filaments in the filopodia of the secondary mesenchyme cells of *Arbacia punctulata* and *Echinarachnius parma. J. Cell Sci.* **5,** 195–210. *96*

Timell, T. E. (1973). Ultrastructure of the dormant and active cambial zones and the dormant phloem associated with formation of normal and compression woods in *Picea abies* (L.) Karst. Syracuse: Pub. 96, Office of Public Service and Continuing Education, State Univ. of NY College of Environmental Science and Forestry. *54, 78*

Todd, G. W. (1972). Water deficits and enzymatic activities. In: *Water deficits and plant growth,* vol. 3, ed. T. T. Kozlowski, pp. 177–216. New York: Academic Press. *98*

Torrey, J. G. (1953). The effect of certain metabolic inhibitors on vascular tissue differentiation in isolated pea roots. *Am. J. Bot.* **40,** 525–33. *36*

Torrey, J. G. (1963). Cellular patterns in developing roots. *Symp. Soc. Exp. Biol.* **17,** 285–314. *9, 66*

Torrey, J. G. (1966). The initiation of organized development in plants. In: *Advances in*

Morphogenesis, vol. 5, ed. M. Abercrombie & J. Brachet, pp. 39–91. New York: Academic Press. *4, 8, 9, 64*

Torrey, J. G. (1967). Morphogenesis in relation to chromosomal constitution in long-term plant tissue cultures. *Physiol. Plant.* **20,** 265–75. *49*

Torrey, J. G. (1968). Hormonal control of cytodifferentiation in agar and cell suspension cultures. In: *Biochemistry and physiology of plant growth substances,* ed. F. Wightman & G. Setterfield, pp. 843–55. Ottawa: Runge Press. *7, 20*

Torrey, J.G. (1971). Cytodifferentiation in plant cell and tissue cultures. In: *Les cultures de tissus de plantes.* Colloques Internationaux CNRS no. 193, pp. 177–86. Paris:CNRS. *8, 24, 114*

Torrey, J. G. & Fosket, D. E. (1970). Cell division in relation to cytodifferentiation in cultured pea root segments. *Am. J. Bot.* **57,** 1072–80. *20, 22, 48*

Torrey, J. G., Fosket, D. E. & Hepler, P. K. (1971). Xylem formation: a paradigm of cytodifferentiation in higher plants. *Am. Sci.* **59,** 338–52. *1, 3, 8, 15, 19, 23, 29, 64, 74, 91, 94*

Torrey, J. G. & Loomis, R. S. (1967*a*). Auxin-cytokinin control of secondary vascular tissue formation in isolated roots of *Raphanus. Am. J. Bot.* **54,** 1098–1106. *66*

Torrey, J. G. & Loomis, R. S. (1967*b*). Ontogenetic studies of vascular cambium formation in excised roots of *Raphanus sativus* L. *Phytomorphology* **17,** 401–9. *66*

Tschermak-Woess, E. (1960). Über den Einbau von ^{3}H-Thymidin in die DNS und die Endomitotischetätigkeiten der Wurzel von *Vicia faba. Chromosoma* **11,** 25–8. *49*

Tulecke, W. (1967). Plastid function in plant tissue cultures. I. Porphyrin synthesis by dark-grown haploid and diploid albino cultures. *Am. J. Bot.* **54,** 797–804. *49*

Ursprung, H. (ed.) (1969). *The stability of the differentiated state in tissue culture.* Berlin: Springer-Verlag. *45*

Van Fleet, D. S. (1962). Histochemistry of enzymes in plant tissues. In: *Handbuch der Histochemie,* vol. 7(2), ed. W. Graumann & K. Newmann, pp. 1–38. Stuttgart: Gustav Fisher Verlag. *88*

van Lith-Vroom, M. L., Gottenbos, J. J. & Karstens, W. K. H. (1960). General appearance, growth pattern and anatomical structure of crown-gall tissue of *Nicotiana tabacum* L. grown *in vitro* on culture media containing glucose or soluble starch as a carbon source. *Acta Bot. Neerl.* **9,** 275–85. *7*

Van Parijs, R. & Vandendriessche, L. (1966). Changes of the DNA content of nuclei during the process of cell elongation in plants. II. Variations other than doublings, of the amount of Feulgen stain in maturing plant cells. *Arch. Int. Physiol. Biochim.* **74,** 587–91. *52*

Van't Hof, J. (1966). Experimental control of DNA synthesizing and dividing cells in excised root tips of *Pisum. Am. J. Bot.* **53,** 970–6. *42*

Van't Hof, J. (1967). Recovery enhancement of G1 and G2 meristematic cells in excised pea roots during an induced post irradiation stationary phase. *Radiat. Res.* **32,** 792–803. *42*

Van't Hof, J. (1968*a*). Control of cell progression through the mitotic cycle by carbohydrate provision. I. Regulation of cell division in excised plant tissue. *J. Cell Biol.* **37,** 733–8. *42*

Van't Hof, J. (1968*b*). The action of IAA and kinetin on the mitotic cycle of proliferative and stationary phase excised root meristems. *Exp. Cell Res.* **51,** 167–76. *47*

Van't Hof, J. (1971). The principal points of control in the mitotic cycle of pea meristem cells; energy considerations, characterization and radiosensitivity. In: *Proc. Fourth International Congress on Radiation Research,* ed. H. Duplan. London: Gordon & Breach (in press). *42*

Van't Hof, J. (1974). Control of the cell cycle in higher plants. In: *Cell cycle controls,* ed. G. M. Padilla, I. L. Cameron & A. M. Zimmermann, pp. 77–85. New York: Academic Press. *40*

Van't Hof, J. & Kovacs, C. J. (1972). Mitotic cycle regulation in the meristem of cultured roots: the principal control point hypothesis. In: *The dynamics of meristem cell populations,* ed. M. W. Miller & C. C. Kuehnert, pp. 15–32. New York: Plenum Press. *42, 43*

Verma, D. P. S. & van Huystee. R. B. (1970). Cellular differentiation and peroxidase isozymes in cell culture of peanut cotyledons. *Can. J. Bot.* **48,** 429–31. *86*

Vigil, E. R. & Ruddat, M. (1973). Effect of gibberellic acid and actinomycin D on the formation and distribution of rough endoplasmic reticulum in barley aleurone cells. *Plant Physiol.* **51,** 549–58. *29*

Vöchting, H. (1892). Über Transplantation am Pflanzenkörper. Untersuchungen zur Physiologie und Pathologie. Tübingen: H. Laupp. *4*

Waller, G. R. & Burström, H. (1969). Diterpenoid alkaloids as plant growth inhibitors. *Nature,* (Lond.) **222,** 576–8. *30*

Wangermann, E.(1970). Autoradiographic localization of soluble and insoluble ^{14}C from (^{14}C)-indolylacetic acid supplied to isolated *Coleus* internodes. *New Phytol.* **69,** 919–27. *85*

Wardlaw, C. W. (1947). Experimental investigations of the shoot apex of *Dryopteris aristata* Druce. *Phil. Trans. Roy. Soc. Ser. B* **232,** 343–84. *69*

Wardrop, A. B. & Bland, D. E. (1959). The process of lignification in woody plants. In: *Biochemistry of wood,* ed. K. Kratzl & G. Billek, pp. 92–114. New York: Pergamon Press. *82, 88*

Wareing, P. F. (1951). Growth studies in woody species. IV. The initiation of cambial activity in ring-porous species. *Physiol. Plant.* **4,** 546–62. *56, 57*

Wareing, P. F. (1958). Interaction between indoleacetic acid and gibberellic acid in cambial activity. *Nature,* (Lond.) **181,** 1744–5. *8, 61*

Wareing, P. F. (1971). Some aspects of differentiation in plants. In: *Control mechanisms of growth and differentiation,* ed. D. D. Davies & M. J. Balls, pp. 323–44. London: Cambridge University Press. *11, 12, 45*

Wareing, P. F., Haney, C. E. A. & Digby, J. (1964). The role of endogenous hormones in cambial activity and xylem differentiation. In: *The formation of wood in forest trees, ed. M. H. Zimmermann, pp. 323–*44. New York: Academic Press. *8, 57,61*

Webb, J. L. (1963). *Enzyme and metabolic inhibitors,* vol. 1. New York: Academic Press. *106*

Webb, J. L. (1966). *Enzyme and metabolic inhibitors,* vols. 2, 3. New York: Academic Press. *106*

Webster, B. D. & Radin, J. W. (1972). Growth and development of cultured radish roots. *Am. J. Bot.* **59,** 744–51. *66*

Webster, P. L. & Van't Hof, J. (1970). DNA synthesis and mitosis in meristems: requirements for RNA and protein synthesis. *Am. J. Bot.* **57,** 130–9. *42*

Wessells, N. K. (1964). DNA synthesis, mitosis, and differentiation in pancreatic acinar cells *in vitro. J. Cell Biol.* **20,** 415–33. *46*

Wetmore, R. H. & Rier, J. P. (1963). Experimental induction of vascular tissues in callus of angiosperms. *Am. J. Bot.* **50,** 418–30. *7, 64, 89, 91, 114*

Wetmore, R. H. & Sorokin, S. (1955). On the differentiation of xylem. *J. Arnold Arbor.* **36,** 305–24. *7*

Whiting, A. C. & Murray, M. A. (1946). Histological responses of bean plants to phenylacetic acid. *Bot. Gaz.* **107,** 312–32. *7*

Whitmore, F. W. (1968). Auxin and cell wall formation in the cambial derivatives of pine and cottonwood. *Forest Sci.* **14,** 197–205. *58*

Whitmore, F. W. & Jones, B. M. (1972). Altered auxin-gibberellin ratios in abnormal vascular development of peach. *Plant Physiol.* **49**(suppl.), 32. *62*

Whitmore, F. W. & Zahner, R. (1966). Development of the xylem ring in stems of young pine trees. *Forest Sci.* **12,** 198–210. *63*

Wilbur, F. H. & Riopel, J. L. (1971*a*). The role of cell interaction on the growth and differentiation of *Pelargonium hororum* cells *in vitro.* I. Cell interaction and growth *Bot. Gaz.* **132,** 183–93. *97, 99*

Wilbur, F. H. & Riopel, J. L. (1971*b*). The role of cell interaction in the growth and diferentiation of *Pelargonium hororum* cells *in vitro.* II. Cell interaction and differentiation. *Bot. Gaz.* **132,** 193–202. *97, 98, 99*

Wilson, B. F. (1963). Increase in cell wall surface area during enlargement of cambial derivatives in *Abies concolor. Am. J. Bot.* **50,** 95–102. *77*

Wilson, B. F., Wodzicki, T. J. & Zahner, R. (1966). Differentiation of cambial derivatives: proposed terminology. *Forest Sci.* **12,** 438–40. *54*

Wimber, D. E. (1963). Methods for studying cell proliferation with emphasis on DNA labels. In: *Cell proliferation,* ed. L. F. Lamerton & R. J. Fry, pp. 1–17. Oxford: Blackwell. *38*

Wimber, D. E. (1966). Duration of the nuclear cycle in *Tradescantia* root tips at three temperatures as measured with ^{3}H-thymidine. *Am. J. Bot.* **53,** 21–4. *39*

Winter, A. (1967). The promotion of the immobilization of auxin in *Avena* coleoptiles by triiodobenzoic acid. *Physiol. Plant.* **20,** 330–6. *107*

Winter, A. (1968). 2,3,5-Triiodobenzoic acid and the transport of 3-indoleacetic acid. In: *Biochemistry and physiology of plant growth substances,* ed. F. Wightman & G. Setterfield, pp. 1063–76. Ottawa: Runge Press. *107*

Witham, F. H. (1968). Effect of 2,4-dichlorophenoxyacetic acid on the cytokinin requirement of soybean cotyledon and tobacco stem pith callus tissues. *Plant Physiol.* **43,** 1455–7. *21*

Wodzicki, T. J. & Humphreys, W. J. (1972). Cytodifferentiation of maturing pine tracheids: the final stage. *Tissue and Cell* **4,** 525–8. *79*

Wodzicki, T. J. & Humphreys, W. J. (1973). Maturing pine tracheids. Organization of intravaculor cytoplasm. *J. Cell Biol.* **56,** 263–5. *79, 81*

Wodzicki, T. J. & Wodzicki, A. B. (1973). Auxin stimulation of cambial activity in *Pinus silvestris.* II. Dependence upon basipetal transport. *Physiol. Plant* **29,** 288–92. *59*

Wood, H. N. & Braun, A. C. (1967). The role of kinetin (6-furfurylaminopurine) in promoting division in cells of *Vinca rosea L. Ann. N.Y. Acad. Sci.* **144,** 244–50. *35*

Wood, H. N. & Braun, A. C. (1973). 8-Bromoadenosine 3′ : 5′-cyclic monophosphate as a promoter of cell division in excised tobacco pith parenchyma tissue. *Proc. Nat. Acad. Sci.* (USA) **70,** 447–50. *35*

Wood, H. N., Lin, M. C. & Braun, A. C. (1972). The inhibition of plant and animal adenosine 3′ :5′-cyclic monophosphate phosphodiesterases by a cell-division-promoting substance from tissues of higher plant species. *Proc. Nat. Acad. Sci.* (USA) **69,** 403–6. *35*

Wooding, F. B. P (1968). Radioautographic and chemical studies of incorporation into sycamore vascular tissue walls. *J. Cell Sci.* **3,** 71–80. *74, 82*

Wooding, F. B. P. (1969). Absorptive cells in protoxylem: association between mitochondria and the plasmalemma. *Planta* **84,** 235–8. *112*

Wooding, F. B. P. & Northcote, D. H. (1964). The development of the secondary wall of the xylem in *Acer pseudoplatanus. J. Cell Biol.* **23,** 327–37. *71*

Wooding, F. B. P. & Northcote, D. H. (1965). An anomalous wall thickening and its possible role in the uptake of stem-fed tritiated glucose by *Pinus pinea. J. Ultrastruct. Res.* **12,** 463–72. *112, 114*

Wright, B. E. (1973). *Critical variables in differentiation*. Englewood Cliffs: Prentice-Hall. *2*

Wright, K. & Bowles, D. J. (1974). Effects of hormones on the polysaccharide-synthesizing membrane systems of lettuce pith. *J. Cell Sci.*, **16,** 433–43. *81*

Wright, K. & Northcote, D. H. (1972). Induced root differentiation in sycamore callus. *J. Cell Sci.* **11,** 319–37. *90, 91*

Wright, K. & Northcote, D. H. (1973). Differences in ploidy and degree of intercellular contact in differentiating and non-differentiating sycamore calluses. *J. Cell Sci.* **12,** 37–53. *22, 25–6, 50, 96, 97*

Wunderlich, F., Müller, R. & Speth, V. (1973). Direct evidence for a colchicine-induced impairment in the mobility of membrane components. *Science* **182,** 1136–8. *72*

Yeoman, M. M. & Aitchison, P. A. (1973). Changes in enzyme activities during the division cycle of cultured plant cells. In: *The cell cycle in development and differentiation*, ed. M. Balls & F. S. Billett, pp. 185–201. London: Cambridge University Press. *40, 84*

Yeoman, M. M. & Brown, R. (1971). Effects of mechanical stress on the plane of cell division in developing callus cultures. *Ann. Bot.* **35,** 1101–12. *102*

Yeoman, M. M. & Davidson, A. W. (1971). Effect of light on cell division in developing callus cultures. *Ann. Bot.* **35,** 1085–1100. *34, 100*

Yeoman, M. M. & Evans. P. K. (1967). Growth and differentiation of plant tissue cultures, II. Synchronous cell divisions in developing callus cultures. *Ann. Bot.* **31,** 323–32. *49*

Yeung, E. C. & Peterson, R. L. (1974). Ontogeny of xylem transfer cells in *Hieracium floribundum*. *Protoplasma* **80,** 155–74. *112, 113, 114*

Young, B. S. (1954). The effect of leaf primordia on differentiation in the stem. *New Phytol.* **53,** 445–60. *12*

Young, C. W. & Hodas, S. (1964). Hyroxyurea: inhibitory effect on DNA metabolism. *Science* **146,** 1172–4. *40*

Young, C. W., Schochetman, G. & Karnofsky, D. A. (1967). Hydroxyurea-induced inhibition of deoxyribonucleotide synthesis: studies on intact cells. *Cancer Res.* **27,** 526–34. *40*

Zajaczkowski, S. (1973). Auxin stimulation of cambial activity in *Pinus silvestris*. I. The differential cambial response. *Physiol. Plant.* **29,** 281–7. *8, 57, 59*

Ziegler, H. (1970). Morphactins. *Endeavor* **29,** 112–16. *109, 110*

Ziegler, H. & Sankhla, N. (1974). *Morphactins*. New York: Academic Press (in press). *109*

Zimmerman, W. A. (1936). Untersuchungen über die räumliche und zietliche Verteilung des Wuchsstoffes bei Bäumen. *Z. Bot.* **30,** 209–52. *8, 59*

Zimmermann, M. H. & Brown, C. L. (1971). *Trees: structure and function*. New York: Springer-Verlag. *3, 55, 57, 63*

Zobel, R. W. (1973). Some physiological characteristics of the ethylene-requiring tomato mutant diageotropica. *Plant Physiol.* **52,** 385–9. *8, 32*

Zobel, R. W. (1974). Control of morphogenesis in the ethylene-requiring tomato mutant, diageotropica. *Can. J. Bot.* **52,** 735–41. *32, 33, 35*

Note added in proof. Van't Hof, J. (1971) citation should read:

Van't Hof, J. (1973). Two principal points of control in the mitotic cycle of pea meristem cells: energy considerations, characterization, and radiosensitivity. In: *Advances in Radiation Research. Biology & Medicine*, vol. 2, ed. J. F. Duplan & A. Chapiro, pp. 881–94. London: Gordon & Breach.

Subject index